THE UNIVERSAL MANDELBROT SET

Beginning of the Story

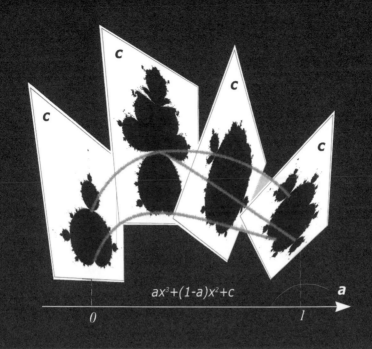

$ax^3+(1-a)x^2+c$

THE UNIVERSAL MANDELBROT SET

Beginning of the Story

V Dolotin
A Morozov

ITEP, Russia

 **World Scientific**

NEW JERSEY · LONDON · SINGAPORE · BEIJING · SHANGHAI · HONG KONG · TAIPEI · CHENNAI

Published by

World Scientific Publishing Co. Pte. Ltd.

5 Toh Tuck Link, Singapore 596224

USA office: 27 Warren Street, Suite 401-402, Hackensack, NJ 07601

UK office: 57 Shelton Street, Covent Garden, London WC2H 9HE

Library of Congress Cataloging-in-Publication Data
Dolotin, V. (Valery)
    The universal mandelbrot set : beginning of the story / V. Dolotin & A. Yu. Morozov
    p. cm.
    Includes bibliographical references.
    ISBN 981-256-837-9 (alk. paper)
    1. Mandelbrot set. I. Title. II. Morozov, A. Yu. (Alexei Yurievich), 1961–

QA614.86 .D65 2006
514'.742--dc22   2006049629

British Library Cataloguing-in-Publication Data
A catalogue record for this book is available from the British Library.

Printed in Singapore by World Scientific Printers (S) Pte Ltd

Preface

This book is devoted to the structure of Mandelbrot set – a remarkable and important personage of the modern theoretical physics, related to chaos and fractals and simultaneously to analytical functions, Riemann surfaces and phase transitions. This makes Mandelbrot set one of the bridges connecting the worlds of chaos and order (integrability). At the same time the Mandelbrot set is a very simple object, allowing detailed *experimental* investigation with the help of the easily available computer programs and profound *theoretical* analysis with the help of undergraduate-level mathematics. This makes Mandelbrot set a wonderful subject of physical investigation: one needs to look for theoretical explanations of experimentally observable properties of the object and theoretical predictions can be immediately verified by the easily affordable new experiments. In this book we just lift the cover of Pandora box: if interested, the reader can easily continue along any of the research lines, which are mentioned below, or, digging a little deeper into the experimental data, find and resolve a lot of new puzzles, – all this without any special mathematical background. Remarkably, when thinking about this simple subject one appears in the close vicinity of modern problems of the "high theory": the subject is not only simple, it is also deep.

Mandelbrot set shows up in the study of trajectories (world lines) behavior change under a change of the laws of motion. Generally speaking, trajectory depends on two types of data: on the motion law (e.g. on the choice of the Hamiltonian) and on the initial (boundary) conditions. The pattern of motion (the phase portrait of the system) can be rather sophisticated: trajectories can tend to fixed points, to limit cycles, to strange attractors and go away from the similar types of structures, some trajectories can go to infinity; trajectories can wrap around each other, knot and unknot etc. In this book the information about the phase portrait will be collected in **Julia sets**, the structure of these sets encodes information about the world lines dependence on the initial conditions. If the law of motion is changed, the phase portrait = Julia set is also deformed. These changes can be pure quantitative (due to the variations of the shape of trajectories, of foci and limit cycles positions) and – sometime – qualitative (looking as reshuffling of the phase portrait, creation or elimination of the fixed points, cycles, attractors etc.). The **Mandelbrot set** or its **boundary**, to be precise, is exactly the set of all the points in the laws-of-motion space (say in the space of Hamiltonians) where the qualitative reshufflings (phase transitions) take place. Therefore the problems about the Mandelbrot set (or, better, about the set of the phase portraits, associated with various laws of motion, i.e. about the **"sheaf of Julia sets over the Mandelbrot set"**) are the typical problems about the structure of the **space of theories** (laws of motion), about what changes and what remains intact in transition from one theory to another. In other words, these are typical problems from the *string theory*[1] – the science which studies the various sets of similar physical models and studies the change of the experimentally measurable characteristics with the change of the model. Moreover, in the case of the Mandelbrot set one can immediately address the problems from a difficult branch of string theory, which nowadays attracts a good deal of attention, – from *landscape theory*, which wonders *how often* is a given property encountered in the given set of models (i.e. associates with the various physical quantities, say, scattering amplitudes or mass spectra, the *measures* on the space of theories, which specify how frequent are the amplitudes of the given type or masses of the given values in the models of the given type). One of the lessons, taught by the study of Mandelbrot sets, is that even if the **phase transitions** are relatively rare in the space of theories, the phase transition points are not distributed in a uniform way, they tend to condense and form clusters of higher dimensions, up to codimension one, which actually separate – the naively un-separated – domains in the space of theories and make this space disconnected. This lesson is important for the theory of effective actions, for problems of their analytical continuations, for construction of special functions (τ-functions)

---

[1] As often happens, the name is pure historical and refers to a concrete set of models, which were first successfully approached from such direction. Better name could be "the theory of theories" (sometime "abbreviated" to pretentious "theory of everything") or, most adequate, "generic quantum field theory". Unfortunately, these names are already associated with narrower fields of research, where accents are put on somewhat different issues.

which can describe these effective actions, and for many other – conceptual and practical – applications of string theory. Dynamics, essential for this type of applications of Mandelbrot-set theory, is not obligatory the ordinary dynamics in physical space-time, often the relevant one is dynamics in the space of coupling constants, i.e. the (generalized) renormalization group flow.

In this book we restrict consideration to *discrete* dynamics *of a single variable*. This restriction preserves most essential properties of the subject, but drastically simplifies computer simulations and mathematical formalism (for example, substitutes the study of arbitrary analytic functions by that of polynomials or the study of arbitrary spectral surfaces of arbitrary dimension by that of the ordinary Riemann surfaces). However, throughout the text we use every chance to show directions of possible generalizations. Getting rid of the infinite-dimensional algebras (loop algebras), which would be unavoidable in considerations of continuous dynamics, we concentrate instead on *another* infinity: unification of various one-dimensional Mandelbrot sets, shown in numerous pictures below and associated with particular 1-dimensional families of evolution operators, into a total (infinite-dimensional) **universal Mandelbrot set**, shown symbolically in Fig. 7.1 at the very end of the book. Understanding the structure of this set (which has a pure cathegorial nature of the **universal discriminant variety**) can be one of the intermediate targets of the theory. The notions of *discriminant* and *resultant* are assumed familiar from the courses on general algebra and are actively used already in the Introduction. Still, if necessary, all the necessary definitions and properties can be found in Sec. 4.12 of the present book (the section itself is devoted to generalizations from polynomials to arbitrary analytic functions).

Acknowledgements. Our work is supported by Federal Program of the Russian Ministry for Industry, Science and Technology No 40.052.1.1.1112, by the Grant of Support for the Scientific Schools 8004.2006.2, NWO project 047.011.2004.026, INTAS project "Current topics in String Theory", ANR-05-BLAN-0029-01 project "Geometry and Integrability in Mathematical Physics" , and by RFBR grants 04-02-17227 (V.D.) and 04-02-16880 (A.M.).

Contents

Chapter 1

Introduction

The theory of deterministic dynamical systems [Arnold(1989)],

$$\dot{x}^i = \beta^i(x), \tag{1.1}$$

is one of the main chapters of theoretical and mathematical physics. One of the goals of the theory is to classify the types of motion – from completely integrable to fully ergodic, – and understand their dependence on the choice of the vector field $\beta(x)$ and initial conditions. Discrete dynamics, in the avatar of Poincaré map,

$$x^i \longrightarrow f^i(x),$$
$$\beta^i(f(x)) = \frac{\partial f^i}{\partial x^j} \beta^j(x), \tag{1.2}$$

naturally arises in attempts to develop such a theory and plays the principal role in computer experiments (which – in the absence of adequate theory – remain the main source of information about dynamical systems).

In quantum field/string theory (QFT) the most important (but by no means unique) appearance of the problem (1.1) is in the role of renormalization group (RG) flow. As usual, QFT relates (1.1) to a problem, formulated in terms of effective actions, this time to a Callan-Symanzik equation for a function $\mathcal{F}(x, \varphi)$ of the *coupling constants* x^i and additional variable φ (having the meaning of logarithm of background *field*)

$$\beta^i(x) \frac{\partial \mathcal{F}}{\partial x^i} + \frac{\partial \mathcal{F}}{\partial \varphi} = 0. \tag{1.3}$$

In other contexts solutions to this equation are known as *characteristics* [1] of the problem (1.1): in the case of one variable the solution is arbitrary function of $\varphi - s(x)$ or of $x(s = \varphi)$, its φ dependence is induced by the action of operator $\exp\left(-\varphi \sum_i \beta^i \partial_i\right)$. Until recently only simple types of motions (attraction and repulsion points) were considered in the context

1

of RG theory, but today this restriction is no longer fashionable (see, for example, [2] for the first examples of RG with periodic behaviour and [3]-[5] for further generalizations). In this context the discrete dynamics is associated with the original Kadanoff-style formulation of renormalization group, and Callan-Symanzik equation (1.3) becomes a finite-difference equation.[1] Bifurcations of dynamical behaviour of trajectories with the change of $\beta(x)$ in (1.1) are associated with transitions between different branches of effective actions [6]. One of the most intriguing points here is that the theory of effective actions is actually the one of *integrable systems* – effective actions are usually τ-functions, RG flows are multi-directional (since the boundary of integration domain in functional integral can be varied in many different ways) and related to Whitham dynamics [7]. Obvious existence of nontrivial RG flows (describing, for example, how a fractal variety changes with the change of scale) – along with many other pieces of evidence – implies existence of a deep relation between the worlds of order and chaos (represented by τ-functions and dynamical systems respectively and related by the – yet underdeveloped – theory of effective actions). Description of effective actions (prepotentials) $\mathcal{F}(\varphi, x)$ in situations when (1.1) exhibits chaotic motion remains an important open problem. Of course, one does not expect to express such $\mathcal{F}$ through ordinary elementary or special functions, but we are going to claim in this book that the adequate language

---

[1]This discrete equation looks like

$$\mathcal{F}_\pm(f^{\circ n}(x), \varphi \pm n) = \mathcal{F}_\pm(x, \varphi)$$

Its generic solution is arbitrary function of any of its particular solutions. Formally, such particular solutions are given by the differences $\varphi \mp \hat{n}(x)$, where $\hat{n}(x)$ is the "time", needed to reach the point x, i.e. a solution of the equation $f^{\circ n}(x_0) = x$ for given f and x_0. In other words, particular solutions are formally provided by the φ-th image and φ-th preimage of $f(x)$,

$$\mathcal{F}_\pm(x, \varphi) = f^{\circ(\mp \varphi)}(x)$$

For example, if $f(x) = x^c$ and $f^\circ(x) = x^{c^n}$, then the generic solution is an arbitrary function of the variable $c^{\mp \varphi} \log x$.

Consideration of the present book provides peculiar φ-independent solutions (zero-modes), associated with periodic f-orbits. For example, the "orbit δ-function"

$$\mathcal{F}_\pm(x, \varphi) = F_n'(x)\delta(F_n(x))$$

is a solution of such type (see Eq. (4.4) below) – a generalization of continuous-case zero-mode

$$\mathcal{F}(x, \varphi) = \det\left(\frac{\partial \beta}{\partial x}\right)\delta(\beta(x)).$$

At least in this sense periodic f-orbits are discrete analogues of non-trivial zeroes of β-function, in continuum limit they can turn into limit cycles and strange attractors.

can still be searched for in the framework of algebraic geometry, in which the theory of dynamical systems can be naturally embedded, at least in the discrete case.

The goal of this book is to overview the connection between the *discrete* dynamics of one *complex* variable [8], which studies the properties of the iterated map

$$x \to f(x) \to f^{\circ 2}(x) := f(f(x)) \to \ldots \to f^{\circ n}(x) \to \ldots, \qquad (1.4)$$

and the algebraic properties of the (locally) analytic function $f(x) = \sum_k a_k x^k$, considered as an element of the (infinite-dimensional) space $\mathcal{M}$ of all analytic functions $\mathbf{C} \to \mathbf{C}$ on the complex plane $\mathbf{C}$. We attribute the apparent complexity of the Julia and Mandelbrot sets, associated with the pattern of orbit bifurcations, to the properties of well defined (but not yet well-studied) algebraic subspaces in moduli space: discriminant variety $\mathcal{D}$ and spaces of shifted n-iterated maps $\mathcal{M}_{on}$

$$F_n(x; f) = f^{\circ n}(x) - x. \qquad (1.5)$$

In other words, we claim that Julia and Mandelbrot sets can be defined as objects in algebraic geometry and studied by theoretical methods, not only by computer experiments.

We remind, that the boundary of Julia set $\partial J(f)$ for a given f is the union of all *unstable* periodic orbits of f in C, i.e. a subset in the union $\mathcal{O}(f)$ of all periodic orbits, perhaps, *complex*, which, in turn, is nothing but the set of all roots of the iterated maps $F_n(x; f)$ for all n. The boundary of *algebraic* Mandelbrot set $\partial M(\mu) \subset \mu \subset \mathcal{M}$ consists of all functions f from a given family $\mu \subset \mathcal{M}$, where the stability of orbits changes. However, there is no one-to-one correspondence between Mandelbrot sets and bifurcations of Julia sets, since the latter can also occur when something happens to *unstable* orbits and some of such events do not necessarily involve *stable* orbits, controlled by the Mandelbrot set. In order to get rid of the reference to somewhat subtle notion of stability (which also involves *real* instead of *complex-analytic* constraints), we suggest to reveal the algebraic structure of Julia and Mandelbrot sets and then use it as a definition of their algebraic counterparts. It is obvious that boundaries of Mandelbrot sets are related to discriminants, $\partial M = \mathcal{D} \subset \mathcal{M}$ and can actually be defined without reference to stability. The same is true also for Julia sets, which can be defined in terms of *grand orbits* of f, which characterize not only the images of points after the action of f, but also their pre-images. The bifurcations of so defined algebraic Julia sets occur at the boundary of what we call the *grand Mandelbrot set*, a straightforward extension of Mandelbrot set, also representable as discriminant variety. Its boundary consists of points in the space of maps (*moduli space*), where the structure of periodic orbits (not obligatory *stable*) and their pre-orbits changes. Hypothetically algebraic Julia sets coincide with the ordinary ones (at least for some appropriate choice of stability criterium).

To investigate the algebraic structure of Julia and Mandelbrot sets we suggest to reformulate the problem in terms of representation theory.

An analytic function as a map $f : \mathbf{C} \to \mathbf{C}$ defines an action of $\mathbf{Z}$ on $\mathbf{C}$: $n \mapsto \underbrace{f \circ \ldots \circ f}_{n}$. For f with real coefficients this action has two invariant subsets $\mathbf{R}$ and $\mathbf{C} - \mathbf{R}$. Often in discrete dynamics one considers only the action on $\mathbf{R}$. However, some peculiar properties of this action (like period doubling, stability etc.) become transparent only if we consider it as a special case of the action of f with complex coefficients and no distinguished subsets (like $\mathbf{R}$) in $\mathbf{C}$.

The set of periodic orbits of this action can be considered as a representation of Abelian group $\mathbf{Z}$, which depends on the shape of f. One can study these representations, starting from generic representation, which corresponds to generic position of f in the space $\mathcal{M}$. Other representations result from merging of the orbits of generic representation, what happens when f belongs to certain discriminant subsets $\mathcal{D}_n^* = (\mathcal{D} \cap \mathcal{M}_n)^* \subset \mathcal{M}$. These discriminant sets are algebraic and have grading, related to the order of the corresponding merging orbits. The boundary of Mandelbrot set $\partial M(\mu)$ for a one-parametric family of maps $\mu \subset \mathcal{M}$ is related to the union of the countable family of algebraic sets of increasing degree, $\cup_{n=0}^{\infty} (\mathcal{D} \cap \mu_{on})$, and this results in the "fractal" structure of Mandelbrot set. This approach allows us to introduce the notion of discriminant in the (infinite-dimensional) space $\mathcal{M}$ of analytic functions as the inverse limit of a sequence of finite-dimensional expressions corresponding to discriminants in the spaces of polynomials with increasing degree.

On the preimages of periodic orbits (on bounded grand orbits) the above $\mathbf{Z}$-action is not invertible. The structure of the set of preimages of a given order k has special bifurcations which are not expressible in terms of bifurcations of periodic orbits. This implies the existence of a hierarchy of discriminant sets in the space of coefficients (secondary Mandelbrot sets). The union of secondary Mandelbrot sets for all k presumably defines all the bifurcations of Julia sets. We call this union the Grand Mandelbrot set.

Mentioned above is just one of many possible "definitions" of Julia and Mandelbrot sets [9]. Their interesting (fractal) properties look universal and not sensitive to particular way they are introduced, which clearly implies that some canonical universal structure stands behind. The claim of the present book is that this canonical structure is algebraic and is nothing but the intersection of iterated maps with the universal discriminant and resultant varieties $\mathcal{D} \subset \mathcal{M}$ and $\mathcal{R} \subset \mathcal{M} \times \mathcal{M}$, i.e. the appropriately defined pull-back $\mathcal{R}^* \subset \mathcal{M}$. Different definitions of Mandelbrot sets are just different vies/projections/sections of $\mathcal{R}^*$. The study of Mandelbrot sets is actually the study of $\mathcal{R}^*$, which is a complicated but well-defined problem of algebraic geometry. Following this line of thinking we suggest to substitute intuitive notions of Julia and Mandelbrot sets by their much better defined *algebraic* counterparts. The details of these definitions can require modifications when one starts proving, say, existence or equivalence theorems, but the advantage of such approach is the very possibility to formulate and prove theorems.

Our main conclusions about the structure of Mandelbrot and Julia sets are collected in Chap. 3. There are no properties of these fractal sets which could not be discovered and explained by pure algebraic methods. However, the adequate part of algebraic theory is not well developed and requires more attention and work. In the absence of developed theory – and even developed language – a lot of facts can be better understood through examples than through general theorems. We present both parts of the story – theory and examples – in parallel.[2] For the sake of convenience we begin the next chapter by listing the main relevant notions and their inter-relations.

---

[2]To investigate examples, we used powerful discriminant and resultant facilities of MAPLE and a wonderful *Fractal Explorer* (FE) program [10]. In particular, the pictures of Mandelbrot and Julia sets for 1-parameter families of maps in this book are generated with the help of this program.

Chapter 2

Notions and notation

2.1 Objects, associated with the space X

• **Phase space X.** In principle, in order to define the Mandelbrot space only **X** needs to be a topological space. However, to make powerful algebraic machinery applicable, **X** should be an algebraically closed field: the complex plane **C** suits all needs, while the real line **R** does not. We usually assume that **X** = **C**, but algebraic construction is easily extendable to p-adic fields and to Galois field. In fact, further generalizations are straightforward: what is really important is the ring structure (to define maps from **X** into itself as polynomials and series) and a kind of Bezout theorem (allowing to parameterize the maps by collections of their roots).[1]

• **A map** f : **X** → **X** can be represented as finite (polynomial) or infinite series

$$f(x) = \sum_{k=0}^{\infty} a_k x^k. \tag{2.1}$$

---

[1]In order to preserve these properties, multidimensional discrete dynamics can be introduced on phase space $\mathbf{X^m}$ in the following way. The x-variables are substituted by m-component vectors $x_1, \ldots, x_m$ (or even $m+1$-component if the space is $\mathbf{CP^m}$ and affine parameterization is used). The relevant maps f : $\mathbf{X^m} \to \mathbf{X^m}$ are defined by tensor coefficients $a_{i;\vec{k}}$,

$$f_i(x) = \sum_{k_1,\ldots,k_m=0}^{\infty} a_{i;k_1,\ldots,k_m} x_1^{k_1} \ldots x_m^{k_m}, \quad i = 1, \ldots, m$$

instead of (2.1). Then one can introduce and study discriminants, resolvents, Mandelbrot and Julia sets just in the same way as we are going to do in one-dimensional situation ($m = 1$). For relevant generalization of discriminant see [11]. Concrete multidimensional examples should be examined by this technique, but this is beyond the scope of the present book.

The series can be assumed convergent at least in some domain of $\mathbf{X}$ (if the notion of convergency is defined on $\mathbf{X}$). Some statements in this book are extendable even to formal series.

- **f-orbit** $O(x; f) \subset \mathbf{X}$ of a point $x \in \mathbf{X}$ is a set of all images,
$$\{x, f(x), f^{\circ 2}(x), \ldots, f^{\circ n}(x), \ldots\}.$$
It is convenient to agree that for $n = 0$ $f^{\circ 0}(x) = x$. The orbit is **periodic** of order n if n is the smallest positive integer for which $f^{\circ n}(x) = x$.

- **Grand f-orbit** $GO(x; f) \subset \mathbf{X}$ of x includes also all the pre-images: all points x', such that $f^{\circ k}(x') = x$ for some k. There can be many different x' for given x and k. The *orbit* is finite, if for some $k \geq 0$ and $n \geq 1$ $f^{\circ (n+k)}(x) = f^{\circ k}(x)$. If k and n are the smallest for which this property holds, we say that x belongs to the k-th preimage of a periodic orbit of order n (which consists of the n points $\{f^{\circ k}(x), \ldots, f^{\circ (n+k-1)}(x)\}$). The corresponding *grand orbit* (i.e. the one which ends up in a finite orbit) is called **bounded grand orbit** (BGO), it looks like a graph, with infinite trees attached to a loop of length n. Grand orbit defines the ramification structure of functional-inverse of the map f and of associated Riemann surface. Different branches[2] of the grand-orbit tree define different branches of the prepotential (the discrete analogue of $\mathcal{F}$ in Eq. (1.3)).

- The union $\mathcal{O}_n(f) \subset \mathbf{X}$ of all periodic f-orbits of order n, the union $\mathcal{O}_{n,k}(f) \subset \mathbf{X}$ of all their k-preimages and the unions $\mathcal{O}(f) = \cup_{n=1}^{\infty} \mathcal{O}_n(f) \subset \mathbf{X}$ of all periodic orbits and $\mathcal{GO}(f) = \cup_{k=0,n=1}^{\infty} \mathcal{O}_{n,k}(f) \subset \mathbf{X}$ of all *bounded* grand orbits. The sets $\mathcal{O}(f) \subset \mathcal{GO}(f)$ can be smaller than $\mathbf{X}$ (since there are also non-periodic orbits and thus unbounded grand orbits).

- The sets $\mathcal{S}_n(f) \subset \mathbf{X}$ and $\mathcal{S}_{n,k}(f) \subset \mathbf{X}$ of all roots of the functions
$$F_n(x; f) = f^{\circ n}(x) - x \tag{2.2}$$
and
$$F_{n,k}(x; f) = f^{\circ (n+k)}(x) - f^{\circ k}(x) = F_{n+k}(x; f) - F_k(x; f) = F_n\left(x; f^{\circ k}\right) \tag{2.3}$$
All periodic orbits $\mathcal{O}_m(f) \subset \mathcal{S}_n(f)$ whenever m is a divisor of n (in particular, when $m = n$). Similarly, $\mathcal{O}_{m,k}(f) \subset \mathcal{S}_{n,k}(f)$. This property allows

---

[2]For a rooted tree, any path, connecting the root and some highest-level vertex is called **branch**. For infinite trees, like most grand orbits, the branches have infinite length.

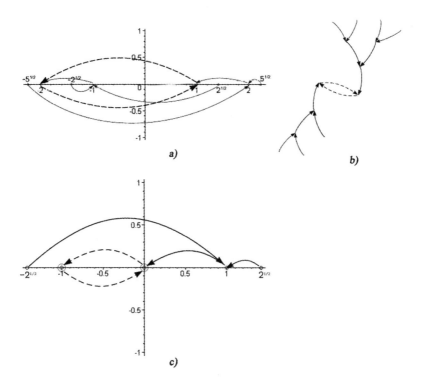

Fig. 2.1 Typical view of a bounded grand orbit of order $n = 2$ (thus the loop is of length 2) for $f(x)$ which is a quadratic polynomial (thus two arrows enter every vertex). (a) Example of generic case: $f(x) = x^2 - 3$, shown together with its embedding into the complex-x plane. (b) The same example, $f(x) = x^2 - 3$, only *internal* tree structure is shown. (c) Example of degenerate case: $f(x) = x^2 - 1$ (only one arrow enters -1, its second preimage is infinitesimally close to 0).

to substitute the study of (bounded) periodic (grand) orbits by the study of roots of maps $F_{n,k}(x; f)$. Reshufflings (mergings and decompositions) of (grand) orbits occur when the roots coincide, i.e. when the map becomes degenerate. This property allows to substitute the study of the phase space $\mathbf{X}$ (where the orbits and roots are living) by the study of the space $\mathcal{M}$ of functions, where the varieties of degenerate functions are known as discriminants $\mathcal{D}$.

- $\mathcal{O}_n(f) \subset \mathbf{X}$ is a set of zeroes of function $G_n(x; f)$, which is an

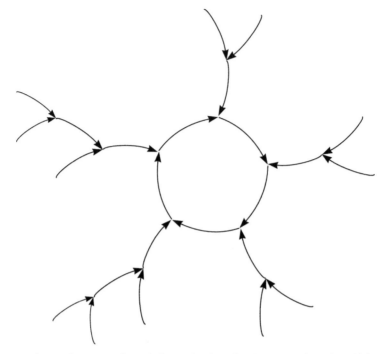

Fig. 2.2   A non-degenerate bounded grand orbit of order $n = 5$ for cubic $f(x)$. Only internal structure is shown, without embedding into **X**.

irreducible-over-any-field divisor of reducible function $F_n(x; f)$,

$$F_n(x; f) = \prod_{m|n} G_m(x; f), \tag{2.4}$$

where the product is over all divisors m of n, including $m = 1$ and $m = n$. The functions G_n are more adequate for our purposes than F_n, but the F_n are much easier to define and deal with in sample calculations.

• **Stable periodic orbit** of order n is defined by conditions

$$\begin{cases} |(f^{\circ n})'(x)| < 1 \text{ i.e. } |F_n'(x) + 1| < 1 \\ G_n(x) = 0 \end{cases} \tag{2.5}$$

for all its points. This definition is self-consistent because

$$(f^{\circ n})'(x) = \prod_{k=0}^{n-1} f'(f^{\circ k}(x)) = \prod_{z \in \text{orbit}} f'(z), \tag{2.6}$$

i.e. the l.h.s. is actually independent of the choice of $x \in O_n(f)$. Other periodic orbits are called **unstable**. Bounded grand orbits are called stable and unstable when periodic orbits at their ends are stable and unstable respectively.

• The set of stable periodic orbits $\mathcal{O}_+(f)$ and its complement: the state of unstable periodic orbits $\mathcal{O}_-(f)$. By definition $\mathcal{O}_+(f) \cup \mathcal{O}_-(f) = \mathcal{O}(f)$.

• **Julia set** $J(f) \in \mathbf{X}$ is attraction domain of stable periodic orbits of f, i.e. the set of points $x \in \mathbf{X}$ with orbits $O(x; f)$, approaching the set of stable orbits: $\forall \varepsilon > 0 \ \exists k$ and $\exists z_k \in \mathcal{O}_+(f)$: $|f^{\circ k}(x) - z_k| < \varepsilon$. This definition refers to non-algebraic structures like continuity. However, the boundary $\partial J(f)$ is an almost algebraic object, since it actually consists of all the unstable periodic orbits and their grand orbits, which are dense in $\partial J(f)$:

$$\partial J(f) = \overline{\mathcal{O}_-(f)} = \overline{\mathcal{G}\mathcal{O}_-(f)}. \qquad (2.7)$$

Moreover, *each* grand orbit "originates in the vicinity" of $\mathcal{O}_-(f)$: generically, every branch of the grand-orbit tree is associated with – "originates at" – a particular orbit from $\mathcal{O}_-(f)$, and when $\mathcal{O}_+(f) = \emptyset$ this is a one-to-one correspondence. If $\mathcal{O}_+(f) \neq \emptyset$, the "future" of the grand orbit can be three-fold: it can terminate in a periodic orbit, stable or unstable, approach (tend to) some stable orbit or go to infinity (which can actually be considered as a "reservoir" of additional stable orbits). If $\mathcal{O}_+(f) = \emptyset$, the situation is different: strange attractors can also occur. Description of $\partial J(f)$ in terms of orbits works even when $\mathcal{O}_+(f) = \emptyset$, but then it is not a boundary of anything, the Julia set $J(f)$ itself is not defined.

For a better description of Julia sets see summary in Chap. 3 and Sec. 5.5.

2.2 Objects, associated with the space $\mathcal{M}$

• The **moduli space** $\mathcal{M}$ of all maps. In principle it can be as big as the set of formal series (2.1), i.e. can be considered as the space of the coefficients $\{a_k\}$. Some pieces of theory, however, require additional structures, accordingly one can restrict $\mathcal{M}$ to sets of continuous, smooth, locally analytic or any other convenient sets of maps. We assume that the maps are single-valued, and their functional inverses can be described in terms of trees, perhaps, of infinite valence. The safest (but clearly over-restricted) choice is a subspace $\mathcal{P} \subset \mathcal{M}$ of all polynomials with complex coefficients, which, in turn, can be decomposed into spaces of polynomials of particular degrees d: $\mathcal{P} = \cup_d^\infty \mathcal{P}_d$. One can also consider much smaller subspaces: families $\mu(\vec{c}) \subset \mathcal{M}$, with $a_k(\vec{c})$ in (2.1) parameterized by several complex parameters $\vec{c} = (c_1, c_2, \ldots)$. Most examples in the literature deal with one-parametric families.

• The moduli subspace $\mathcal{M}_{on} \subset \mathcal{M}$ of *shifted n-iterated maps*, consisting of all maps $F : X \longrightarrow X$ which have the form (2.2) for some $f(x)$. The shift by x is important. Equation (2.2) defines canonical mappings

$$\hat{I}_n : \mathcal{M} \longrightarrow \mathcal{M}_{on},$$

which can also be written in terms of the coefficients: for $f(x)$ parameterized as in Eq. (2.1), and

$$F_n(x; f) = f^{on}(x) - x = \sum_{k=0}^{\infty} a_k^{(n)} x^k$$

we can define $\hat{I}_n$ as an algebraic map

$$\hat{I}_n : \{a_k\} \longrightarrow \{a_k^{(n)}\} \qquad (2.8)$$

where all $a_k^{(n)}$ are polynomials of $\{a_l\}$.[3] Inverse map $\hat{I}_n^{-1}$ is defined on $\mathcal{M}_{on}$ only, not on entire $\mathcal{M}$, but is not single-valued. In what follows we denote the pull-back of $\hat{I}_n$ by *double* star, thus $(\hat{I}_n f)^{**} = f$ (while $\hat{I}_n^{-1}(\hat{I}_n f)$ can contain a set of other functions in addition to f, all mapped into one and the same point $\hat{I}_n f$ of $\mathcal{M}_{on}$).

The map $\hat{I}_n$ can be restricted to polynomials of given degree, then

$$\hat{I}_n : \mathcal{P}_d \longrightarrow \mathcal{P}_{d^n} \qquad (2.9)$$

embeds the $(d+1)_{\mathbf{C}}$-dimensional space into the $(d^n+1)_{\mathbf{C}}$ one as an algebraic variety.[4] Like in (2.9), often $\mu_{on} = \hat{I}_n(\mu) \not\subset \mu$. One can, however, use a

---

[3]In order to avoid confusion we state explicitly that $a_k^{(0)} = 0$ and $a_k^{(1)} = a_k - \delta_{k,1}$. Also $\mathcal{M}_{o1} = \mathcal{M}$, but in general analogous relation is not true for subsets $\mu \subset \mathcal{M}$: μ_{o1} does not obligatory coincide with μ, it differs from it by a shift by x.

[4]For example, $(\mathcal{P}_2)_{o2}$ is embedded into $\mathcal{P}_4$ as algebraic variety:

$$8A^2(D+1) - 4ABC + B^3 = 0,$$
$$64A^2(4AB^2E + B^3 - 4A^2(D+1)^2)^3 = B^3(16A^2(D+1) - B^3)^3$$

This follows from

$$Ax^4 + Bx^3 + Cx^2 + Dx + E = (ax^2 + bx + c)^{o2} - x$$
$$= a^3x^4 + 2a^2bx^3 + (ab^2 + ab + 2a^2c)x^2 + (2abc + b^2 - 1)x + (ac^2 + bc + c).$$

Note that because of the shift by x one should be careful with projectivization of this variety.

pull-back to rectify this situation: $[\mathcal{B} \cap \mu_{\circ n}]^{**} \subset \mu$ for any $\mathcal{B} \subset \mathcal{M}$.

In the studies of *grand orbits* more general moduli subspaces are involved: $\mathcal{M}_{\circ(n,k)} \subset \mathcal{M}$, consisting of all maps of the form (2.3) for some f.

• Similarly, the functions $G_n(x; f)$ define another canonical set of mappings,

$$\hat{J}_n : \mathcal{M} \longrightarrow \mathcal{M}_n$$

into subspaces $\mathcal{M}_n \subset \mathcal{M}$, consisting of all functions which have the form $G_n(x; f)$ for some f. These are, however, somewhat less explicit varieties than $\mathcal{M}_{\circ n}$, because such are the functions $G_n(x; f)$ – defined as irreducible constituents of $F_n(x; f)$. Still, they are quite explicit, say, when f are polynomials of definite degree, and they are more adequate to describe the structures, relevant for discrete dynamics. The J-pullback will be denoted by a single star.

• **Discriminant variety** $\mathcal{D} \subset \mathcal{M}$ consists of all *degenerate* maps f, i.e. such that $f(x)$ and its derivative $f'(x)$ have a common zero. It is defined by the equation

$$D(f) = 0$$

in $\mathcal{M}$, where $D(f)$ is the square of the Van-der-Monde product of roots differences:

$$\text{for } f(x) \sim \prod_k (x - r_k) D(f) \sim \prod_{k \neq l} (r_k - r_l).$$

Restrictions of discriminant onto the spaces of polynomials of given degree, $\mathcal{D}(\mathcal{P}_d) = \mathcal{D} \cap \mathcal{P}_d$, are algebraic varieties in $\mathcal{P}_d$, because $D(f)$ is a polynomial (of degree $2d-1$) of the coefficients $\{a_k\}$ in representation (2.1) of f (while roots themselves are not polynomial in $\{a_k\}$ the square of Van-der-Monde product is, basically, by Vieta's theorem). Discriminant variety $\mathcal{D}$ has singularities of various codimensions k in $\mathcal{D}$, associated with mergings of $k+2$ roots of f.

• **Resultant** variety $\mathcal{R} \subset \mathcal{M} \times \mathcal{M}$ consists of pairs of maps $f(x)$, $g(x)$ which have common zero. It is defined by the equation

$$R(f, g) = 0,$$

and

$$\text{for } f(x) \sim \prod_k (x - r_k) \text{ and } g(x) \sim \prod_l (x - s_l)$$

$$R(f,g) \sim \prod_{k,l} (r_k - s_l).$$

Again, if f and g are polynomials, resultant is a polynomial in coefficients of f and g. Resultant variety has singularities when more than one pair of roots coincide. Discriminant can be considered as appropriate resultants' derivative at diagonal $f = g$. Higher resultant varieties in $\mathcal{M} \times \ldots \times \mathcal{M}$ are also important for our purposes, but will not be considered in the present book.

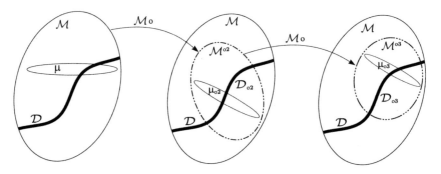

Fig. 2.3   Schematic view of inter-relations between the subsets $\mathcal{M}_{on}$, μ_{on} and $\mathcal{D}_{on}$.

• Intersections of $\mathcal{D}$ and $\mathcal{R}$ with subspaces like $\mathcal{M}_{on}$ and $\mathcal{M}_n$, i.e. $\mathcal{D}_{on} = \mathcal{D} \cap \mathcal{M}_{on}$, $\mathcal{R}_{o(m,n)} = \mathcal{R} \cap (\mathcal{M}_{om} \times \mathcal{M}_{on})$; $\mathcal{D}_n = \mathcal{D} \cap \mathcal{M}_n$, $\mathcal{R}_{m,n} = \mathcal{R} \cap (\mathcal{M}_m \times \mathcal{M}_m)$ consist of disconnected components (see Fig. 2.3). $\hat{I}$-pullbacks $\mathcal{D}_{on}^{**}$, $\mathcal{R}_{o(m,n)}^{**} \subset \mathcal{M}$ and $\hat{J}$-pullbacks

$$\mathcal{D}_n^* = \{f : D(G_n(f)) = 0\} \subset \mathcal{M},$$

$$\mathcal{R}_{m,n}^* = \{f : R(G_m(f), G_n(f)) = 0\} \subset \mathcal{M}$$

also consist of numerous components, disjoint and touching. Particular families $\mu \subset \mathcal{M}$ intersect these singular varieties and provide particular sections $\partial M(\mu)$ of this generic structure: the **universal discriminant** (or resultant) **variety**

$$\mathcal{D}^* = \mathcal{R}^* = \bigcup_{n=1}^{\infty} \mathcal{R}_n = \overline{\bigcup_{n,m=1}^{\infty} \mathcal{R}_{m,n}^* \bigcup_{n=1}^{\infty} \mathcal{D}_n^*}$$

which is the boundary of the **Universal Mandelbrot set** $\partial M(\mathcal{M})$.

- For a given family of maps $\mu \subset \mathcal{M}$ the **boundary of** *algebraic* **Mandelbrot set** $\partial M(\mu) \subset \mu$ is defined as a $\hat{J}$-pullback

$$\partial M(\mu) = \cup_{n=1}^{\infty}(\mathcal{D}_n^* \cap \mu) = (\mathcal{D} \cap (\cup_{n=1}^{\infty}\mu_n))^*$$
$$= \mu \cup (\cup_{n=1}^{\infty}\{f : \ D(G_n(f)) = 0\}) \tag{2.10}$$

- Varieties $\mathcal{D}^{**}$, $\mathcal{R}^{**}$, $\mathcal{D}^*$, $\mathcal{R}^*$ – and thus the boundary of algebraic Mandelbrot space ∂M – can instead be considered as pure topological objects in $\mathcal{M}$ independent of any additional algebraic structures, needed to define F_n, G_n, discriminant and resultant varieties $\mathcal{D}$ and $\mathcal{R}$. For example, $\mathcal{D}_1^{**} = \mathcal{D}_1^*$ consists of all maps $f \in \mathcal{M}$ with two coincident fixed points. Higher components $\mathcal{D}_n^* \subset \mathcal{D}_n^{**}$ consist of maps f, with two coincident fixed points of their n-th iteration (for $\mathcal{D}_n^{**}$) and such that any lower iteration does not have coincident points (for $\mathcal{D}_n^*$). Resultants consist of maps f with coinciding fixed points of their different iterations. Fixed points and iterated maps are pure categorial notions, while to define coincident points one can make use of topological structure: two different fixed points merge under continuous deformation of f in $\mathcal{M}$.

- **Stability domain** of the periodic order-n orbits $S_n \subset \mathcal{M}$ is defined by the system (2.5): if a root x of $G_n(f)$ is substituted into inequality, it becomes a restriction on the shape of f, i.e. defines a domain in $\mathcal{M}$. This domain is highly singular and disconnected. For a family $\mu \subset \mathcal{M}$ we get a section $S_n(\mu) = S_n \cap \mu$, also singular and disconnected.
- **Mandelbrot set** $M(\mathcal{M}) \subset \mathcal{M}$ is a union of all stability domains with different n:

$$M = \cup_{n=1}^{\infty}S_n, \quad M(\mu) = \cup_{n=1}^{\infty}S_n(\mu) = M \cap \mu. \tag{2.11}$$

Definitions (2.11) and (2.10) leave obscure most of the structure of Mandelbrot set and its boundary. Moreover, even consistency of these two definitions is not obvious.[5] For better description of Mandelbrot set see Chap. 3.

---

[5]It deserves emphasizing that from algebraic perspective there is an essential difference between the Mandelbrot and Julia sets themselves and their boundaries. Boundaries are pure algebraic (or, alternatively, pure topological) objects, while entire spaces depend on stability criteria, which, for example, break complex analyticity and other nice properties present in the description of the boundaries.

• Above definitions, at least the algebraic part of the story, can be straightforwardly extended from orbits to grand orbits and from maps $F_n(f)$ and $G_n(f)$ to $F_{n,k}(f)$ and $G_{n,k}(f)$. The maps $G_{n,k}(x; f)$, which define k-th pre-orbits of periodic order-n orbits are not obligatory *irreducible* constituents of $F_{n,k}(x; f)$ (which is the case for $k = 0$). They are instead related to peculiar map

$$z: \mathcal{M} \to \mathbf{X}$$

associating with every $f \in \mathcal{M}$ the points $x \in \mathbf{X}$ with degenerate pre-image:

$$z_f = \{z: \ D(f(x) - z) = 0\}.$$

Generalizations of Mandelbrot sets to $k \geq 1$ are called **secondary Mandelbrot sets**. The *Grand Mandelbrot set* is the union of secondary sets with all $k \geq 0$. Bifurcations of *Julia sets* with the variation of f inside $\mu \subset \mathcal{M}$ are captured by the structure of Grand Mandelbrot set.

We see that, the future theory of dynamical systems should include the study of two purely algebro-geometric objects:

(i) the universal discriminant and resultant varieties $\mathcal{D} \subset \mathcal{M}$ and $\mathcal{R} \subset \mathcal{M} \times \mathcal{M}$,

(ii) the map $\mu \longrightarrow \breve{\mu} = \cup_{n=1}^{\infty} \mu_n$ and the pull-backs $\mathcal{D}^* \subset \mathcal{M}$ and $\mathcal{R}^* \subset \mathcal{M}$.

It should investigate intersections of $\mathcal{D}$ and $\breve{\mu}$ and $\mathcal{R}$ with $\breve{\mu} \times \breve{\mu}$ and consider further generalizations: to multi-dimensional phase spaces $\mathbf{X}$ and to continuous iteration numbers n. It is also interesting to understand how far one can move with description of $\mathcal{R}^*$ and universal Mandelbrot space by pure topological methods, without the ring and other auxiliary algebraic structures.

2.3   Combinatorial objects

• **Divisors tree** $t[n]$ is a finite rooted tree. Number n stands at the root, and it is connected to all $\tau(n) - 1$ of its divisors $k|n$ (including $k = 1$, but excluding $k = n$). Each vertex k is further connected to all $\tau(k) - 1$ of *its* divisors and so on. The links are labeled by ratios $m = n/k \geq 2$ and $m_i = k_i/k_{i+1} \geq 2$. Number n is equal to product of m's along every branch. See examples in Fig. 2.4. The number $\tau(n)$ of divisors of $n = p_1^{a_1} \dots p_k^{a_k}$ is equal to $\tau(n) = (a_1 + 1) \dots (a_k + 1)$. A generating function (Dirichlet function) has asymptotic

$$D(x) = \sum_{n \leq x} \tau(n) \sim x \log x + (2C - 1) + O(\sqrt{x}), \qquad (2.12)$$

C – Euler constant.

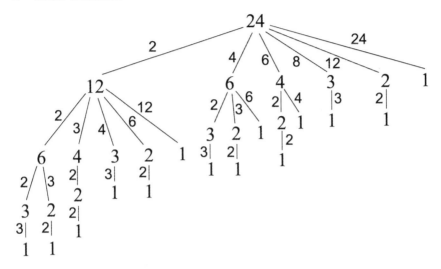

Fig. 2.4 Divisors tree $t[24]$. It contains divisor trees for all its divisors: $t[12]$, $t[6]$, $t[4]$, $t[3]$, $t[2]$, $t[1]$, some in several copies.

• **Multipliers tree** T_n is a rooted tree of infinite valence. At the root stands the number n. The branches at every level p are labeled by positive integers $m_p = 1, 2, 3, 4, \ldots$. Every vertex at p-th level is connected to the root by a path $m_1, m_2, \ldots, m_p$ and the product $n m_1 m_2 \ldots m_p$ stands in it. The **basic forest** $T = \cup_{n=1} T_n$. A number of times $B(n)$ the number n occurs in the forest T is equal to the number of branches in its *divisors tree*.

The numbers $\tau_p(n)$ of ways to represent n as a product of p integers, i.e. the numbers of times n appears at the p-th level of *multipliers tree*, are described by the generating function

$$D_p(x) = \sum_{n \leq x} \tau_p(n) = \frac{1}{2\pi i} \int_{c-i\infty}^{c+i\infty} \frac{ds}{s} x^s \zeta^p(s),$$

$\zeta(s) = \sum_{m=1}^{\infty} m^{-s}$ is Riemann's *zeta*-function. Note that (2.12) describes $\tau(n) = \tau_2(n)$.

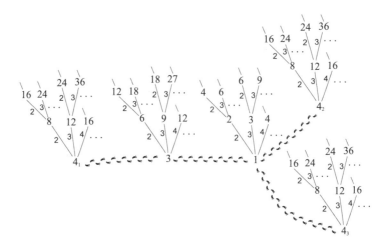

Fig. 2.5   A piece of the forest of multipliers trees. Trails are shown by "steps".

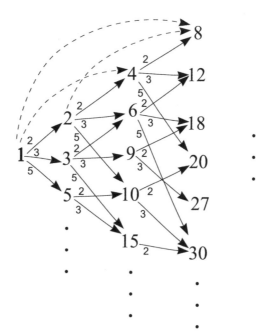

Fig. 2.6.   A fragment of the basic graph. Only the generating arrows, associated with primes, are shown (to avoid overloading the picture). All compositions of arrows should be added as separate arrows (for example, arrows connect 1 to 4, 6, 8, 9 and all other integers).

• **Basic graph** $\mathcal{B}$ is obtained from the multipliers tree T_1 by identification of all vertices with the same numbers, so that the vertices of $\mathcal{B}$ are in one-to-one correspondence with all natural numbers. Since T_1 was a rooted tree, the graph $\mathcal{B}$ is directed: all links are arrows. There are $\tau(n) - 1$ arrows entering the vertex n and infinitely many arrows which exit it and lead to points mn with all natural m.

2.4 Relations between the notions

Relations are summarized in the following table.

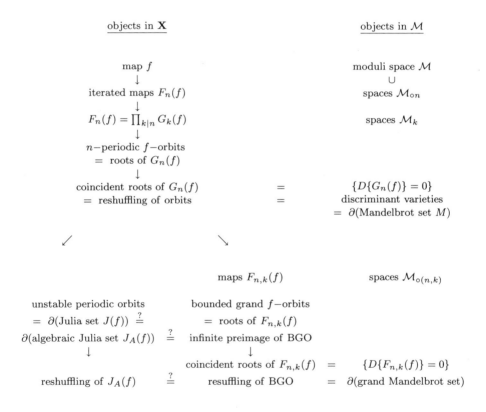

objects in **X**		objects in $\mathcal{M}$
map f		moduli space $\mathcal{M}$
↓		∪
iterated maps $F_n(f)$		spaces $\mathcal{M}_{on}$
↓		
$F_n(f) = \prod_{k\mid n} G_k(f)$		spaces $\mathcal{M}_k$
↓		
$n-$periodic $f-$orbits $=$ roots of $G_n(f)$		
↓		
coincident roots of $G_n(f)$ $=$ reshuffling of orbits	$=$ $=$	$\{D\{G_n(f)\} = 0\}$ discriminant varieties $= \partial$(Mandelbrot set M)

	maps $F_{n,k}(f)$	spaces $\mathcal{M}_{o(n,k)}$
unstable periodic orbits $= \partial$(Julia set $J(f)$) $\overset{?}{=}$ ∂(algebraic Julia set $J_A(f)$) ↓	bounded grand $f-$orbits $=$ roots of $F_{n,k}(f)$ $\overset{?}{=}$ infinite preimage of BGO ↓	
reshuffling of $J_A(f)$	$\overset{?}{=}$ coincident roots of $F_{n,k}(f)$ resuffling of BGO	$=$ $\{D\{F_{n,k}(f)\} = 0\}$ $= \partial$(grand Mandelbrot set)

Chapter 3

Summary

In this chapter we briefly summarize our main claims, concerning the structure of Mandelbrot and Julia sets. They are naturally split into three topics.

3.1 Orbits and grand orbits

This part of the story includes:
- The theory of orbit and pre-orbit functions $G_n(x; f)$ and $G_{n,k}(x; f)$.
- Classification of orbits, pre-orbits and grand orbits.
- Intersections (bifurcations) of orbits and degenerations of grand orbits.
- Discriminant and resultant analysis, reduced discriminants d_n and resultants $r_{m,n}$ intersections of resultant and discriminant varieties.

All these subjects – to different level of depth – are considered in Chap. 4 below and illustrated by examples in Chaps. 5 and 6. Systematic theory is still lacking.

3.2 Mandelbrot sets

Figures 3.1(a-b) show three hierarchical levels of Mandelbrot set: the structure of individual component (M_1 in Fig. 3.1(a)); existence of infinitely many other components (like $M_{3,\alpha}$ in Fig. 3.1(b), which look *practically* the same after appropriate rescaling; trails connecting different components $M_{k\alpha}$ with M_1, which are densely populated by other $M_{m\beta}$ (well seen in the same Fig. 3.1(b)). These structures (except for the self-similarity property) are universal: they reflect the structure of the universal Mandelbrot set, of which the boundary is the universal resultant variety $\mathcal{R}^*$. This means that they do not depend on the particular choice of the family μ (above figures are drawn for $f_c = x^4 + c$, which is in no way distinguished, as is obvious from similar pictures for some other families, presented in Chaps. 5 and 6). The only characteristics which depend on the family are the sets of indices α, labeling degeneracies: different domains with the same place in algebraic structures. Algebraic structure reflects intersection properties of different subvarieties $\mathcal{R}^*_{m,n}$, while α's parameterize the intersection subvarieties. To classify the α-parameters one should introduce and study higher resultants and multi-parametric families $\mu \subset \mathcal{M}$, what is straightforward, but (except for introductory examples) is left beyond the scope of the present book.

The main property of the resultant varieties $\mathcal{R}^*_{m,n}$, responsible for the structure of Mandelbrot set, is that they are non-trivial in real codimension two only if n is divisible by m or vice versa. All the rest follows from the above-described relation between the resultant varieties and the boundary of Mandelbrot set. Stability domains S_n and Mandelbrot sets M are made out of the same disk-like[1] building blocks, but connect these blocks in two different ways. Mandelbrot set and its "forest with trails" structure reflect the universal (μ-independent) labeling (ordering) of these **elementary domains** (denoted by σ below), while their sizes, locations are self-similarity properties (which depend on the choice of μ) are dictated by stability equations.

3.2.1   *Forest structure*

For any family of maps μ Mandelbrot set $M(\mu)$ has the following hierarchical structure, see Fig. 3.1(a) (for particular μ some components – or, better to say, their intersections with μ – can be empty):

$M(\mu)$ consists of infinitely many disconnected components (of which only one is clearly seen in Fig. 3.1(a) and some others are revealed by zooming in Fig. 3.1(b)),

$$M(\mu) = \bigcup_{k,\alpha} M_{k\alpha}(\mu), \quad M_{k\alpha} \cap M_{l\beta} = \emptyset, \quad \text{if } k \neq l \text{ or } \alpha \neq \beta,$$

labeled by natural number k and an extra index α, belonging to the μ-dependent set $\nu_k(\mu)$. For polynomial f the set is finite, its size will be denoted by $|\nu_k(\mu)|$.

Every component $M_{k\alpha}(\mu)$ is a union of elementary domains, each with topology of a disc (interior of a $(2D-1)_{\mathbf{R}}$-dimensional sphere if μ is a

---

[1]In the case of multi-parametric families $\mu \subset \mathcal{M}$, when dimension of components of Mandelbrot set $M(\mu)$ is equal to $\dim \mu$, "disc" may not be a very adequate name. In fact, these components look more like interiors of cylinders and tori rather than balls.

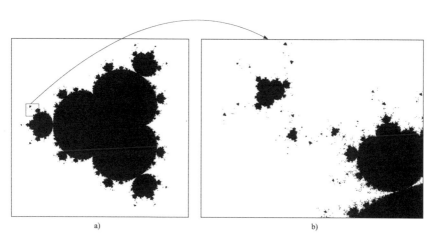

Fig. 3.1   a) General view of the Mandelbrot set for the family $f_c(x) = x^4 + c$. Actually, only the M_1 component is well seen. Arrow points at another component, $M_{3,\alpha}$, shown enlarged in b). The structure of the trail between M_1 and $M_{3,\alpha}$ (almost unobservable in this picture) is also seen in b). Z_3-symmetry of the picture is due to invariance of equations like $x^4 + c = x$ under the transformation $x \to e^{2\pi i/3} x$, $c \to e^{2\pi i/3} c$. Associated Julia set has Z_4 symmetry, see Fig. 3.2. b) Enlarged component $M_{3,\alpha}$, included in the box in a). It *looks* like exact copy of M_1 in a). In fact similarity is approximate, but the deviations are damped by the ratio of sizes, $r_3/r_1 \ll 1$. In this enlarged picture also the trail is seen between M_1 and $M_{3,\alpha}$ and other $M_{k\alpha}$ component in it.

$D_{\mathbf{C}}$-parametric family), which form a tree-like structure. Vertices of the tree $T_{k\alpha}$ – the **skeleton of $M_{k\alpha}$** – are associated with "centers" of elementary domains, a link connects two vertices whenever the two corresponding domains have a common point – and there can be at most one such point, where the two domains actually *touch* each other. The skeleton is a rooted tree, moreover, it is actually a *covering* of the multipliers tree T_k, introduced in Sec. 2.3. The difference is that every branch at the level p carrying the label m_p has multiplicity $|\nu_{m_p}^{(p)}(\mu)|$ and thus an extra label $\alpha_p \in \nu_{m_p}^{(p)}(\mu)$. The union of skeleton trees is the skeleton of the Mandelbrot set – the **Mandelbrot forest**

$$T(M) = \bigcup_{k,\alpha} T_{k\alpha}.$$

Thus

$$M_{k\alpha} = \bigcup_{p=0}^{\infty} M_{k\alpha}^{(p)}.$$

$$M_{k\alpha}^{(p)} = \bigcup_{\substack{m_1,\dots,m_p=1 \\ \alpha_1,\dots,\alpha_p}}^{\infty} \sigma^{(p)} \begin{bmatrix} m_1 & \dots & m_p \\ \alpha_1 & \dots & \alpha_p \end{bmatrix} k\alpha \end{bmatrix}$$

and, finally, $M(\mu) = M \cap \mu$,

$$M = \bigcup_{k,\alpha} \bigcup_{p=0}^{\infty} \left(\bigcup_{\substack{m_1,\dots,m_p=1 \\ \alpha_1,\dots,\alpha_p}}^{\infty} \sigma^{(p)} \begin{bmatrix} m_1 & \dots & m_p \\ \alpha_1 & \dots & \alpha_p \end{bmatrix} k\alpha \end{bmatrix} \right).$$

The elementary domains are $\sigma^{(p)} \begin{bmatrix} m_1 & \dots & m_p \\ \alpha_1 & \dots & \alpha_p \end{bmatrix} k\alpha \end{bmatrix}$ and two of them touch at exactly one point, whenever they are connected by a link in the powerful tree $T_{k\alpha}$, i.e. when $p' = p + 1$, and $m_i' = m_i$, $\alpha_i' = \alpha_i$ for all $i = 1, \dots, p$.

3.2.2   *Relation to resultants and discriminants*

The touching point (map $f \in \mu$) belongs to the resultant variety $\mathcal{R}_{n,n'}^*$, i.e. is a root of $R(G_n(f), G_{n'}(f)) = 0$, with $n = km_1 \dots m_p$ and $n' = km_1 \dots m_p m_{p+1}' = nm_{p+1}'$. The boundaries of all elementary domains $\sigma^{(p)}$ with $p > 0$ are smooth (unless μ crosses a singularity of $\mathcal{R}$), while $\partial\sigma^{(0)}[k\alpha]$ has a cusp, located at discriminant variety $\mathcal{D}_k^*$, i.e. at $f \in \mu$, satisfying $D(G_k(f)) = 0$ ($\sigma^{(0)}[k\alpha]$ itself has a peculiar cardioid form). It is clear that the bigger the family μ, the more intersections it has with the resultant and discriminant varieties, thus the bigger are the sizes of the sets $\nu^{(p)}(\mu)$. For the entire $\mathcal{M}$ the indices α_p get continuous and parameterize the entire pullbacks of the resultant and discriminant varieties.

Every elementary domain $\sigma^{(p)}$ touches a single domain of the lower level $p-1$ and infinitely many domains of the next level $p+1$ (which are labeled by all integers $m = m_{p+1} \geq 2$ and $\alpha = \alpha_{p+1} \in \nu_m^{(p+1)}(\mu)$). The touching points – belonging to zeroes of $R(G_n, G_{nm})$ with all possible $m = m_{p+1}$ are actually dense in the boundary $\partial\sigma^{(p)}$, i.e. the boundary can be considered as a closure of the sets of zeroes.

3.2.3   *Relation to stability domains*

As already mentioned, every $\sigma^{(p)} \begin{bmatrix} m_1 & \dots & m_p \\ \alpha_1 & \dots & \alpha_p \end{bmatrix} k\alpha \end{bmatrix}$ is characterized by an integer $n = km_1 \dots m_p$. Sometime, when other parameters are inessential, we even denote it by $\sigma_n^{(p)}$. Stability domain $S_n(\mu)$, defined by the conditions (2.5) is a union of all $\sigma(\mu)$ with the same n. This implies, that S_n is actually a sum over all branches of *divisor tree* $t[n]$, introduced in Sec. 2.3. Then p, the length of the branch at k, is its end-link. Other links carry numbers $m_1, \ldots, m_p$. The only new ingredient is the addition of extra parameters α at every step. This means that we actually need α-decorated divisor trees (for every family μ), which we denote $\tilde{t}[n](\mu)$ and imply that links at p-th level carry pairs m_p, α_p with $\alpha_p \in \nu_{m_p}^{(p)}(\mu)$, and sum over decorated trees imply summation over α's. In other words, the elementary domains σ are actually labeled by branches of the decorated trees, $B\tilde{t}$, and

$$S_n = \bigcup_{B\tilde{t}[n]} \sigma_n[B\tilde{t}] = \bigcup_{p=0}^{\infty} \bigcup_{\substack{k|n \\ \alpha \in \nu_k}} \left(\bigcup_{\substack{m_1,\ldots,m_p: \ n=km_1\ldots m_p \\ \alpha_1 \in \nu_{m_1}^{(1)}, \ldots, \alpha_p \in \nu_{m_p}^{(p)}}} \sigma^{(p)} \left[\begin{array}{ccc} m_1 & \ldots & m_p \\ \alpha_1 & \ldots & \alpha_p \end{array} \middle| k\alpha \right] \right).$$

(3.1)

It remains an interesting problem to prove *directly* that Eqs. (2.5) imply decomposition (3.1).

3.2.4 *Critical points and locations of elementary domains*

Let w_f denote critical points of the map $f(x)$:

$$f'(w_f) = 0.$$

Then $(f^{\circ n})'(w_f) = 0$, i.e. the critical points always belong to some *stable* orbit, see (2.5). This orbit is periodic of order n provided

$$G_n(w_f; f) = 0. \tag{3.2}$$

This is an equation on the map f and its solutions define points f in the family $\mu \subset \mathcal{M}$, which lie inside stability domain S_n, and, actually, inside certain elementary domains $\sigma^{(p)} \left[\begin{array}{ccc} m_1 & \ldots & m_p \\ \alpha_1 & \ldots & \alpha_p \end{array} \middle| k\alpha \right]$. In fact this is a one-to-one correspondence: solutions to Eq. (3.2) enumerate all the elementary domains.

This implies the following pure algebraic description of the powerful forest of $M(\mu)$ (formed by the coverings $T_{k\alpha}$ of the multiplier trees T_k). All vertices of the graph are labeled by solutions of (3.2) and accordingly carry indices n. Links are labeled by roots of the resultants $R(G_m(f), G_n(f))$, with appropriate m and n (standing at the ends of the link). α-parameters serve to enumerate different solutions of (3.2) and different roots of the same resultants.

Critical points w_f also play a special role in the study of bifurcations of grand orbits and Julia sets, see Sec. 4.6.

3.2.5   *Perturbation theory and approximate self-similarity of Mandelbrot set*

For concrete families $\mu \subset \mathcal{M}$ a kind of approximated perturbation theory can be developed in the vicinity of solutions $f = f_0$ to (3.2). Namely, one can expand equations (2.5), which define the shape of stability domain (and thus of the elementary domains $\sigma[k\alpha]$), in the small vicinity of the point $(x, f) = (w_{f_0}, f_0)$ and, assuming that the deviation is small, substitute original exact equations by their approximation for small deviations. Approximate equations have a *universal* form, depending only on the symmetry of the problem. This explains why in many examples (where this method is accurate enough) all the components $M_{k\alpha}$ of Mandelbrot set *look* approximately the same, i.e. why the Mandelbrot set is approximately self-similar (see Fig. 3.1(b)) and help to classify the components of stability domain. The same method can be used to investigate the *shape* (not just structure) of Julia sets. See in Sec. 4.9.4 an example of application of this procedure to the map families $f_c = x^d + c$.

3.2.6   *Trails in the forest*

The last level of hierarchy in the structure of Mandelbrot set is represented by **trails**, see Fig. 3.1(b). Despite the components $M_{k\alpha}$ do not intersect (do not have common points), they are linked by a tree-like system of trails: each $M_{k\alpha}$ is connected to M_1 by a single trail $\tau_{k\alpha}$, which is densely populated by some other components $M_{l\beta}$. The trail structure is exhaustively described by triangle **embedding matrix**, $\tau = \{\tau[k\alpha, l\beta]\}$ with unit entry when one trail is inside another, $\tau_{l\beta} \subset \tau_{k\alpha}$, and zero entry otherwise. If locations of all $M_{k\alpha}$ in parameter-$\vec{c}$ space are known (for example, evaluated with the help of the perturbation theory from Sec. 3.2.5), then embedding matrix fully describes the trails. It is unclear whether any equations in parameter space μ can be written, which define the shape of trails.

Embedding matrix seems to be universal, i.e. does not depend on the choice of the family μ. If true, this means that the trail structure is indeed a pertinent characteristic of Mandelbrot/resultant variety, not of its section by μ. However, as usual, the universal structure is partly hidden because of the presence of non-universal degeneracies, labeled by α-parameters (at our level of consideration, ignoring higher resultants): concrete trail systems $\tau(\mu)$ are *coverings* of presumably-universal system.

Trail structure requires further investigation and does not get much attention in the present book.

3.3   Sheaf of Julia sets over moduli space

The structure of Julia set $J(f)$ is also hierarchical and it depends on the position of the map f in Mandelbrot set, especially on its location in Mandelbrot forest, i.e. on the set of parameters $\{k\alpha, p, m_i\alpha_i\}$, labeling the elementary domain $\sigma^{(p)} \begin{bmatrix} m_1 \ \ldots \ m_p \\ \alpha_1 \ \ldots \ \alpha_p \end{bmatrix} k\alpha$ which contains f. Since in this book we do not classify α-dependencies, only the numbers k (labeling connected component), p (labeling the level) and $m_1, \ldots, m_p$ will be interpreted in terms of $J(f)$ structure.

Julia set is a continuous deformation of a set of discs (balls), which – the set – depends on k; increase of p by every unit causes gluing of infinitely many points at the boundary, at every point exactly m_p points are glued together. Every elementary domain is associated with a set of stable orbits (there is often just one), contained *inside* the Julia set together with their grand orbits. All other (i.e. unstable) periodic orbits and their grand orbits belong to the boundary of Julia set and almost each particular grand orbit fills this boundary densely. At the touching point between two adjacent elementary domains the stable orbit approaches the boundary ∂J from inside J (perhaps, it is better to say that the boundary deforms and some of its points – by groups – approach the orbit which lies *inside J*), intersect some unstable orbit and "exchange stability" with it. Their grand orbits also intersect. The ratio of orders of these two orbits is m_p and every m_p points of unstable *grand* orbit merge with every point of the stable one, thus strapping the disc at infinitely many places (at all points of the merging grand orbits) and pushing its sectors into bubble-like shoots, see Fig. 3.2. On this picture a mixture of merging *pairs* and *triples* is clearly seen. The bridges for pairs became needles, while triplets are touching. Angles between triplets will increase as c goes inside $\sigma_6^{(2)}$ and reach $2\pi/3$ at its boundary, and so on. Arrows point the positions of Julia sets in the Mandelbrot set. The "Julia sheaf" is obtained by "hanging" the corresponding Julia set over each point of the Mandelbrot set. At the boundary of Mandelbrot set the Julia sets are reshuffled and the task of the theory is to describe this entire variety (the sheaf) and all its properties, both for the universal Mandelbrot set and multi-dimensional Julia sets associated with multi-component maps, and for the particular sections, like the one-parametric family of single-component quartic maps, $f(x; c) = x^4 + c$ shown in this particular picture.

As f transfers to the elementary domain at level $p + 1$, the new stable orbit (the one of bigger order) quits the boundary and immerses into Julia set, while the orbit of smaller order, which is now unstable, remains at its boundary: the singularities, created when the grand orbits crossed, cannot disappear.

One can say that every link of the forest describes a phase transition of Julia set between neighboring elementary domains. An order parameter of this transition is the distance between approaching stable and unstable orbits on one side, looking as the contact length between emerging components of Julia set, and the angle between these components on the other side, when they are already separated except for a single common point. Both quantities vanish at transition point (i.e. when f is at the touching point between two domains in the Mandelbrot set). If we move around the transition point in $\mathcal{M}$, there is a non-trivial monodromy (the Julia set gets twisted), but on the way one should obligatory pass through the complement of Mandelbrot set, where the discs – which form Julia set – disappear, only the boundary remains, Julia set has no "body", only boundary, and exact definition of monodromy is hard to give.

Julia sets can change structure not only when stable orbit intersects with unstable one – what happens when the corresponding elementary domains σ of Mandelbrot set touch each other – but also when unstable grand orbits at the boundary $\partial J(f)$ cross or degenerate. Such events are not reflected in the Mandelbrot set itself: a bigger **Grand Mandelbrot set**, which contains all zeroes of $D(F_{n,k})$ and $R(F_{n,k}, F_{m,l})$ on its boundary, should be introduced to capture all the bifurcations of Julia set.

See Sec. 5.5 for more details about Julia sets and their bifurcations.

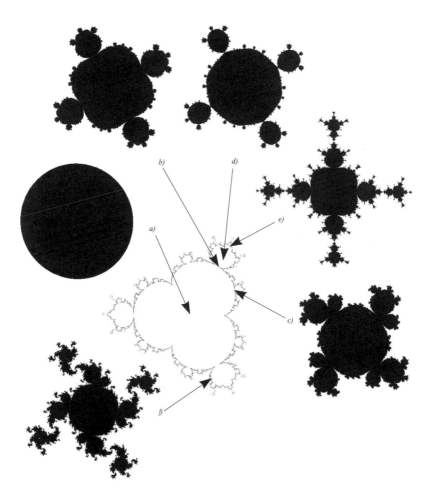

Fig. 3.2 The typical forms of the Julia set for the map $f_c = x^4 + c$ and $c \in M_1$. a) c in the center of $\sigma_1^{(0)}$ ($c = 0$). Z_4 symmetry of the picture is due to the symmetry of the map f_c. Associated Mandelbrot set has symmetry Z_3, see Fig. 3.1(a). b) c at the touching point $\sigma_1^{(0)} \cup \sigma_2^{(1)}$ ($c = \frac{5}{16} 2^{1/3}(1 + i\sqrt{3})$). Infinitely many *pairs* of points on the unit circle (a) are identified to provide this pattern. c) c at another touching point $\sigma_1^{(0)} \cup \sigma_3^{(1)}$ ($c = \frac{1}{8}(9 + i\,111\sqrt{3})^{1/3}$). Infinitely many *triples* of points on the unit circle (a) are identified to provide this pattern. d) c inside $\sigma_2^{(1)}$ ($c = 0.45 + i\,0.8$). This is a further deformation of (a) in the direction of (b). Once appeared in (b) the merging points at the boundary and straps of the disc caused by this merging of boundary points do not disappear, but angles between the external bubbles increase from the zero value which they have in (b). They will turn into $2\pi/2$ and bridges become needles when c reaches the boundary of $\sigma_2^{(1)}$, say an intersection point $\sigma_2^{(1)} \cup \sigma_4^{(2)}$ (e) or $\sigma_2^{(1)} \cup \sigma_6^{(2)}$ (f). e) c at the touching point $\sigma_2^{(1)} \cup \sigma_4^{(2)}$ ($c \approx 0.56 + 0.97i$). f) c at the touching point $\sigma_2^{(1)} \cup \sigma_6^{(2)}$ ($c \approx 0.309 - 0.862i$).

Chapter 4

Fragments of theory

4.1 Orbits and reduction theory of iterated maps

In discrete dynamics trajectories are substituted by *orbits* and closed trajectories – by periodic orbits O_n of the *finite* order n. Different orbits have no common points, and f acts on each O_n by cyclic permutation of points. All points of every O_n belong to the set $\mathcal{S}_n(f)$ of roots of the function

$$F_n(x; f) = f^{\circ n}(x) - x.$$

The map f generates the action of cyclic group $\hat{f}(n)$ on $\mathcal{S}_n(f)$, and this set is decomposed into orbits of $\hat{f}(n)$ of the orders k, which are divisors of n (all O_n with $k = n$ are among them). If n is divisible by k, then $F_n(x)$ is divisible by $F_k(x)$, and in general F_n decomposes into irreducible constituents:

$$F_n(x; f) = \prod_{k|n}^{\tau(n)} G_k(x; f) \tag{4.1}$$

with a product over all possible $\tau(n)$ divisors k of n (including $k = 1$ and $k = n$). The number of periodic orbits of order n is equal to $N_n(f)/n$, where $N_n(f)$ is the number of roots of $G_n(x; f)$. Irreducibility means that G_k are not decomposable into simpler constituents in an **X**- and f-independent way (of course, when $G_k(x)$ are polynomials they can be decomposed into products of monomials over **C**, but this decomposition will not take place over **R**).

Among the tasks of the theory is the study of reducibility of the sets $\mathcal{S}_n(f)$, the ways they decompose into orbits and the ways this decomposition changes under the deformations of f within $\mu \subset \mathcal{M}$. Since the story is essentially about the roots of functions, it gets much more transparent in

31

the *complex* setting than in the *real* one. Also, the entire theory is naturally a generalization from the case of the polynomial functions $f \in \mathcal{P} \subset \mathcal{M}$.

Obviously, for $f \in \mathcal{P}_d \subset \mathcal{P}$, $F_n \in \mathcal{P}_{d^n}$, and $G_k(x)$ in (4.1) is a polynomial of certain degree $N_k(d)$. If $d = p$ is prime, then the group $\hat{f}(n) = \mathbf{Z_n}$ and is isomorphic to the Galois group over F_p of the cyclotomic polynomial $x^{p^n} - x$ (while the Galois group of f itself over $\mathbf{C}$ is trivial). If instead $n = p$ is prime, then d out of the d^p roots of $F_p(x)$ are also the roots of $F_1(x)$, i.e. are invariant points (orbits of order one) of f, while the remaining $d^p - d$ roots decompose into $\frac{d^p - d}{p}$ orbits of order $n = p$. According to (4.1) the numbers $N_n(d)$ for $f \in \mathcal{P}_d$ can be defined recursively: from

$$\sum_{k|n}^{\tau(n)} N_k(d) = \deg F_n = d^n$$

it follows that

$$N_n(d) = d^n - \sum_{\substack{k|n \\ k<n}} N_k(d).$$

The lowest numbers $N_n(d)$ are:

$$N_1(d) = d,$$
$$N_2(d) = d^2 - d = d(d-1),$$
$$N_3(d) = d^3 - d = (d-1)d(d+1),$$
$$N_4(d) = d^4 - d^2 = d^2(d^2-1) = (d-1)d^2(d+1),$$
$$N_5(d) = d^5 - d,$$
$$N_6(d) = (d-1)d(d+1)(d^3+d-1),$$
$$N_7(d) = d^7 - d,$$
$$N_8(d) = d^8 - d^4 = d^4(d^4-1),$$
$$N_9(d) = d^9 - d^3 = d^3(d^6-1),$$
$$N_{10}(d) = d(d^4-1)(d^5+d-1),$$

$$\dots$$

For n prime, $n = p$, $N_p(d) = d^p - d = d(d^{p-1} - 1)$, and small Fermat theorem guarantees that $N_p(d)$ is divisible by p. Further,

$$N_{p^k}(d) = d^{p^k} - d^{p^{k-1}} = d^{p^{k-1}} \left(d^{(p-1)p^{k-1}} - 1 \right),$$
$$N_{p_1 p_2}(d) = d^{p_1 p_2} - d^{p_1} - d^{p_2} + d,$$

in particular,

$$N_{2p}(d) = (d^p - d)(d^p + d - 1),$$
$$N_{3p}(d) = (d^p - d)(d^{2p} + d^{p+1} + d^2 - 1).$$

4.2 Bifurcations and discriminants: from real to complex

Usually considerable part of discrete dynamics concentrates on the *real* part of above story, i.e. deals with the functions f, which map the real line $\mathbf{R} \subset \mathbf{C}$ into itself. Such functions form a subspace $\mathcal{M}^{(\mathbf{R})}$ in $\mathcal{M}$. The subset $\mathcal{P}^{(\mathbf{R})} \subset \mathcal{M}^{(\mathbf{R})}$ consists of all polynomials with real coefficients. In variance with the complex situation, different polynomials F_n in the same $\mathcal{P}_{n^d}^{(\mathbf{R})}$ can have different numbers of *real* roots, thus even the size of the set $\mathcal{S}_n^{(R)}(f)$ can change when f is varying inside $\mathcal{M}^{(R)}$. In fact, the orbits of $f \in \mathcal{M}^{(R)}$ are either entirely real, $O \subset \mathbf{R}$ or entirely complex, $O \subset \mathbf{C} - \mathbf{R}$. In the latter case the orbit is either self-conjugate or there is a complex conjugate orbit $\bar{O} \neq O$. The number of real roots of $F_n(x)$, i.e. the size of the set $\mathcal{S}_n^{(\mathbf{R})}(f)$, can change whenever a self-conjugate orbit or a pair of conjugate orbits, that was complex, becomes real or vice versa, a real orbit turns into a complex self-adjoint or a pair of conjugate orbits. This is generalization of the well-known phenomenon, when a pair of complex-conjugate *roots* of a polynomial from $\mathcal{P}^{(\mathbf{R})}$ becomes real, or vice versa, when a pair of real roots collide and go away into the complex domain – just in the case of peculiar polynomials $F_n(x)$ from $\mathcal{M}_{on}$ this happens at once with entire orbits, not just with *pairs* of roots. In the case of polynomial's roots, the points in $\mathcal{P}^{(\mathbf{R})}$, where some roots can migrate between the real line and complex domain, belong to *discriminant* varieties $\mathcal{D}_d \subset \mathcal{P}_d$, consisting of polynomials with at least a pair of coincident roots (i.e. such that $P(x)$ and $P'(x)$ have at least one common root).[1] The same is true for the orbits of f: all kinds of reshuffling of orbits take place when $F_n \in \mathcal{D} \subset \mathcal{M}$. Note that in the last statement neither the discriminant varieties nor the maps (and their orbits) are restricted to *polynomials* with *real* coefficients, everything is equally well defined at least for the entire space of *complex* polynomials $\mathcal{P}$. Moreover, a generalization should exist to the entire space $\mathcal{M}$ of analytic

---

[1] The simplest example of non-trivial discriminant variety is $\mathcal{D}_2$ – the quadric $b^2 - 4ac = 0$ in the space $\mathcal{P}_2$ of quadratic polynomials $ax^2 + bx + c$. In accordance with (4.34) below

$$b^2 - 4ac = -\frac{1}{a} \det \begin{pmatrix} a & b & c \\ 2a & b & 0 \\ 0 & 2a & b \end{pmatrix}$$

If quadratic polynomial is considered as a quadric in $\mathbf{CP}^1$, $Q(x,y) = ax^2 + bxy + cy^2$, discriminant is a determinant of 2×2 matrix with two lines formed by the coefficients of $\frac{\partial Q}{\partial x}$ and $\frac{\partial Q}{\partial y}$:

$$b^2 - 4ac = -\det \begin{pmatrix} 2a & b \\ b & 2c \end{pmatrix}.$$

(not obligatory polynomial) functions, see Sec. 4.10 below.

Coefficients $a_k^{(n)}$ of

$$F_n(x) = f^{\circ n}(x) - x = \sum_{k=0}^{\infty} a_k^{(n)} x^k$$

are polynomials of the coefficients a_k of $f(x) = \sum_{k=0}^{\infty} a_k x^k$, thus we have an algebraic map (2.8) between the spaces of coefficients $\hat{I}_n : \{a\} \to \{a^{(n)}(a)\}$. Then the space $\mathcal{M}$ of coefficients $\{a\}$ of f is divided into connected components by the $\hat{I}_n$-pullback of the discriminant variety $\mathcal{D}_{on}$. Given a path $a(t)$ in the space $\mathcal{M}$, its image $\hat{I}_n(a(t))$ can cross $\mathcal{D}_{on} = \mathcal{D} \cap \mathcal{M}_{on}$ at a generic (non-singular) point $\hat{I}_n(a(t_0))$. $F_n(x;t)$ will have different number of *real* roots before and after the crossing (i.e. for $t' < t_0 < t''$ in a neighborhood of t_0). This change in the number of real roots is called **bifurcation** in the theory of dynamical systems. Let the number of real roots m' for $t' < t_0$ be less than the number m'' for $t'' > t_0$. If a new root $x_i(t'')$ belongs to an f-orbit O, then *all* the elements of O belong to the f-invariant domain $\mathbf{C} - \mathbf{R}$ for $t' < t_0$ and "simultaneously come" to the f-invariant domain $\mathbf{R}$ at $t = t_0$. Since elements of O and its conjugate $\bar{O}$ come to the real axis $\mathbf{R}$ simultaneously, one can say, that real roots "are born in pairs" at $t = t_0$.

4.3   Discriminants and resultants for iterated maps

General comments on the definitions of discriminants and resultants (including the non-polynomials case) are collected in Sec. 4.11 below. However, for iterated maps these quantities are highly reducible.

Discriminant variety is defined by the equation

$$D(f) = 0$$

where, for polynomial f, $D(f)$ is the polynomial of the coefficients $\{a\}$, see Eq. (4.32) below. Similarly, $D(F_n)$ is a polynomial of the coefficients $\{a^{(n)}(a)\}$ and the resultant $R(F_n, F_m)$ is a polynomial of coefficients $\{a^{(n)}(a)\}$ and $\{a^{(m)}(a)\}$.

Since function $F_n(x)$ is reducible over any field (e.g. over $\mathbf{R}$ as well as over $\mathbf{C}$), see (4.1), the resultant factorization rule (4.35) implies:

$$D(F_n) = \prod_{k|n} D(G_k) \prod_{\substack{k,l|n \\ k>l}} R^2(G_k, G_l). \tag{4.2}$$

However, this is only the beginning of the story. Despite $G_k(x)$ being irreducible constituents of $F_n(x)$, the resultants $R(G_k, G_{k/m})$ and discriminants $D(G_k)$ are still reducible:

$$R(G_k, G_l) = r^l(G_k, G_l), \quad l < k,$$

$$D(G_k) = d^k(G_k) \prod_{m>1}^{n} R^{m-1}(G_k, G_{k/m}) = d^k(G_k) \prod_{l|k} r^{k-l}(G_k, G_l),$$

$$\text{thus} \quad D(F_n) = \prod_{k|n} \left(d^k(G_k) \prod_{l|k} r^{k+l}(G_k, G_l) \right). (4.3)$$

Indeed, whenever a resultant vanishes, a root of some $G_{k/m}$ coincides with that of G_k and then – since roots of G_k form orbits of order k – the k roots should merge by groups of m into $l = k/m$ roots of $G_{k/m}$, and there are exactly l such groups. In other words, whenever a group of m roots of G_k merges with a root of $G_{k/m}$, so do $l = k/m$ other groups, and $R(G_k, G_{k/m})$ is an l-th power of an irreducible quantity, named $r(G_k, G_{k/m})$ is (4.3).

In a little more detail, if $\alpha_1, \ldots, \alpha_k$ with $k = lm$ are roots of G_k and $\beta_1, \ldots, \beta_l$ are those of $G_{k/m}$, then $\alpha_1 = \beta_1$ implies, say, that

$$\alpha_1 = \ldots = \alpha_m = \beta_1,$$

$$\alpha_{m+1} = \ldots = \alpha_{2m} = \beta_2,$$

$$\ldots$$

$$\alpha_{m(l-1)+1} = \ldots \alpha_{lm} = \beta_l.$$

Then in the vicinity of this point

$$R(G_k, G_l) \sim \prod_{i,s}^{l} (\alpha_i - \beta_s) \sim \prod_{s=1}^{l} \left(\prod_{i=(s-1)m+1}^{sm} (\alpha_i - \beta_s) \right)$$

(factors, which do not vanish, are omitted). Internal products are polynomials of the coefficients of f with *first*-order zeroes (the roots themselves are not polynomial in these coefficients!), and only external product enters our calculus and provides power l in the first line of Eq. (4.3).

Similarly,

$$D(G_k) \sim \prod_{i<j} (\alpha_i - \alpha_j)^2$$

$$\sim \prod_{r,s=1}^{l}\left(\prod_{\substack{i,j=(s-1)m+1 \\ i<j}}^{sm}(\alpha_i - \alpha_j)^2\right) = \prod_{s=1}^{l}\left(\prod_{i=(s-1)m+1}^{sm}(\alpha_i - \beta_s)\right)^{m-1}$$

what implies that zero of $D(G_k)$ is of order $l(m-1) = k - m$.

The remaining constituent $d(G_k)$ of discriminant $D(G_k)$ describes intersections among order-k orbits. In this case, whenever two points of two orbits coincide, so do – pairwise – the $k-1$ other points. Thus the corresponding zero is of order k, and irreducible quantity $d(G_k)$ enters $D(G_k)$ in k-th power.

For examination of examples in Chaps. 5 and 6 it is convenient to have explicit versions of (4.3) for a few lowest n and k. It is also convenient to use condensed notation:

$$d_k = d(G_k), \qquad r_{kl} = r(G_k, G_l),$$

$r_{kl} = 1$ unless l is divisor of k, $l|k$ (or vice versa, k is divisor of l). Then

k	$D(F_k)$	$D(G_k)$
1	d_1	d_1
2	$d_1 d_2^2 r_{21}^3$	$d_2^2 r_{21}$
3	$d_1 d_3^3 r_{31}^4$	$d_3^3 r_{31}^2$
4	$d_1 d_2^2 d_4^4 r_{41}^5 r_{42}^6 r_{21}^3$	$d_4^4 r_{41}^3 r_{42}^2$
5	$d_1 d_5^5 r_{51}^6$	$d_5^5 r_{51}^4$
6	$d_1 d_2^2 d_3^3 d_6^6 r_{61}^7 r_{62}^8 r_{63}^9 r_{21}^3 r_{31}^4$	$d_6^6 r_{61}^5 r_{62}^4 r_{63}^3$

$\cdots$

Decomposition of discriminant imply that, as f is varied, new roots can emerge in different ways, when different components of $\mathcal{D}^*$ are crossed. If the new roots are born when the component of $R(G_n, G_k)$ is crossed, where the orbits of orders n and k intersect, then they appear at positions of the previously existing roots of G_k. If this happens at real line, then the phenomenon is known as **period doubling** (it is *doubling*, since when more than two new roots occur at the place of one, they are necessarily complex).

4.4   Period-doubling and beyond

The simplest example of orbit reshuffling with the change of the map f is the period-doubling bifurcation [12], which can be described as follows. Let

x_0 be an invariant point of f, i.e. $F_1(x_0) = f(x_0) - x_0 = 0$. Let us see what happens to an infinitesimally close point $x_0 + \epsilon$.

$$f(x_0 + \epsilon) = x_0 + f'(x_0)\epsilon + \dots,$$
$$f^{\circ 2}(x_0 + \epsilon) = x_0 + [f'(x_0)]^2 \epsilon + \dots,$$

$$\dots$$

We see that the necessary condition for $x_0 + \epsilon$ with infinitesimally small but non-vanishing ϵ to be a root of F_1 or F_2 is $f'(x_0) = 1$ or $f'(x_0) = \pm 1$ respectively (i.e. F_1 or F_2 should be degenerate at the point x_0). The period-doubling bifurcation corresponds to the case $f'(x_0) = -1$, i.e. the map f is such that $F_2(x)$, but not $F_1(x)$, becomes degenerate, then in the vicinity of the corresponding stable point (the common zero x_0 of F_2 and F_2') a new orbit of order 2 can emerge. If, more generally, x_0 is a root of some other F_n, $F_n(x_0) = 0$, i.e. describes some (perhaps, reducible, if n is not a simple number) f-orbit of order (period) n, then the same reasoning can be repeated for $f^{\circ n}$ instead of f:

$$F_n(x_0 + \epsilon) = F_n'(x_0)\epsilon + \dots$$

and

$$F_{2n}(x_0 + \epsilon) = F_{2n}'(x_0)\epsilon + \dots.$$

Derivative $F_{2n}'(x_0) = F_n'(x_0)(F_n'(x_0) + 2)$. Period-doubling bifurcation occurs for f with the property that $(f^{\circ n})'(x_0) = -1$ or $F_n'(x_0) = -2$.

The period-doubling bifurcation, though very important, is not the only one possible: new orbits can emerge in other ways as well.

First of all, *doubling* is relevant only in the case of maps with real coefficients. In fully complex situation one can encounter the situations when a higher power $f'(x_0)^k = 1$, $k > 2$ (and all lower powers of $f'(x_0) \neq 1$): then we have the bifurcation when the period increases by a factor of k and the new orbit emerges in the vicinity of original one (which loses stability, but survives).

Second, the new orbits can emerge "sporadically" at "empty places", with no relation to the previously existing orbits and no obvious criterium to warn about their appearance. The only reason for them to occur is the crossing between discriminant variety $\mathcal{D}$ and moduli spaces $\mathcal{M}_{on}$, $\mathcal{M}_n$ of iterated maps or their irreducible constituents.

4.5   Stability and Mandelbrot set

Above consideration of period-doubling bifurcation implies introduction of the notion of "stable orbits" in the following way: the point x of the orbit of order n, i.e. satisfying $G_n(x; f) = 0$, is called "stable" if

$$|(f^{\circ n})'(x)| \leq 1$$

and "unstable" otherwise. Since

$$(f^{\circ n})'(x) = \prod_{k=1}^{n} f'\left(f^{\circ k}(x)\right) = \prod_{\text{all } z\, \in\, \text{orbit}} f'(z) \qquad (4.4)$$

all points of the orbit are simultaneously either stable or unstable.

The **Julia set** $J(f) \subset \mathbf{X}$ is attraction domain of stable periodic orbits in $\mathbf{X}$. Its boundary $\partial J(f)$ consists of all unstable periodic orbits and their grand orbits.

Excluding x from the pair of stability conditions

$$\begin{cases} G_n(x; f) = 0, \\ |F_n'(x; f) + 1| < 1, \end{cases}$$

we obtain a stability domain S_n of the order-n orbits in moduli space $\mathcal{M}$. All zeroes of reduced discriminants $d(G_n)$ and resultants $r(G_n, G_{mn})$ with arbitrary $m \geq 1$ lie at the boundary of S_n. This property is used in our description of **Mandelbrot set** in Chap. 3.

Mandelbrot set describes exchanges of stability between pairs of orbits: at the boundary of Mandelbrot set stable orbits intersect with unstable ones, stable become unstable, while unstable become stable. Since stable orbits lie *inside* Julia set, and unstable ones – on its boundary, this causes reshuffling between interior and the boundary of $J(f)$. Since all this actually happens with entire *grand* orbits, reshuffling involves infinitely many points and looks like a fractalization of the boundary. In fact, this does not exhaust all possible bifurcations of Julia set $J(f)$: they can also be caused by crossings and degenerations of unstable orbits (with no reference to the stable ones). In order to study bifurcations of Julia sets one should take into consideration the *pre-images* $O_{n,s}$ of periodic orbits of f, associated with the roots of the functions

$$F_{n,s}(x) := f^{\circ(n+s)}(x) - f^{\circ s}(x) = F_n\left(f^{\circ s}(x)\right) \qquad (4.5)$$

and study their reducibility properties. Examples in Chaps. 5 and 6 below demonstrate that consideration of orbit's pre-images, discriminants $D(F_{n,s})$ and resultants $R(F_{n,s}, F_{k,r})$ can indeed capture the bifurcations of $J(f)$, which are overlooked by consideration of orbits alone. Presumably, this provides *complete* theory of Julia sets in terms of Grand Mandelbrot set, which characterizes reshufflings of all orbits, stable and unstable, and is fully algebraic, does not refer to additional stability structure.

4.6 Towards the theory of Julia sets

4.6.1 *Grand orbits and algebraic Julia sets*

For a given (algebraically closed) field $\mathbf{X}$ each series $f = \sum_{n=0}^{d} a_n x^n$, $a_n \in \mathbf{X}$, defines a map $\mathbf{X} \xrightarrow{f} \mathbf{X}$ and thus a (pre-)order $f(x) \succ x$ on $\mathbf{X}$. Then the set of points of $\mathbf{X}$ splits into connected components with respect to $\succ$ ("grand orbits"): we say that x_1, x_2 belong to the same **grand orbit**, iff $x_1 \succ x, x_2 \succ x$ for some $x \in \mathbf{X}$, i.e. $f^{\circ n}(x_1) = f^{\circ m}(x_2)$ for some n and m. In particular, for $n = 0, m = 1$ the set of points x_2 satisfying $x_1 = f(x_2)$ is the f-preimage of a given x_1. So a grand orbit for generic point $x \in \mathbf{X}$ can be represented by an oriented tree with the valency of each given vertex x equal to $d + 1$, where d is the number of roots of the equation $f(z) = x$. For a given x the set x^+ of points $x' \succ x$ will be called the **orbit** of x.

However, if $x \in \mathbf{X}$ is a root of some $F_n(x; f)$, then the *orbit* becomes a closed loop of finite length (order) n, and the grand orbit GO is a $d + 1$-valent graph, obtained by gluing vertices at a distance n on a certain totally ordered chain of the general position tree. For a pair of elements x_α, x_β on a periodic orbit of order n we have $O_n = x_\alpha^+ = x_\beta^+$, moreover both $x_\alpha \succ x_\beta$ and $x_\beta \succ x_\alpha$ is true, so for the points of O_n the relation $\succ$ is just a pre-order. But for each $x \in O_n$ there is a subset $x^- \subset GO$, called the **pre-orbit** of x, being a well-ordered tree rooted at x. The set of points $z \in x^-$ for which $f^{\circ s}(z) = x$ will be denoted x^{-s} (not to be confused with a negative power!), so that $x^- = \cup_s x^{-s}$. Each set x^{-s} belongs to the roots of

$$F_{n,s}(x) = F_{n+s}(x) - F_s(x) = f^{\circ(n+s)}(x) - f^{\circ s}(x).$$

Then for each n we have a finite set of grand orbits with n-periodic "bases" O_n. Then we may characterize the initial map f by the structure of this discrete set of data.

In particular, taking $z \in x^-$ we can regard $z^- \subset x^-$ as a sequence of maps we can take its inverse limit, i.e. the set of sequences $\{z_i\}$ where $z_0 = z, z_i = f(z_{i+1})$ (or $z_i \in x^{-i}$). For different z', z'' (which may in general belong to different grand orbits as well) their preorbits $(z')^-, (z'')^-$ are not only isomorphic as ordered sets, but have isomorphic inverse limits. In particular, the set of limit points for $\lim_{\leftarrow}(z')^-$ and $\lim_{\leftarrow}(z'')^-$ in **F** coincide. We call them *the algebraic Julia set* of f:

$$J_A(f) = \lim_{\leftarrow}(x)^- \quad \forall x \in \mathbf{X}.$$

Hypothesis: $J_A(f)$ coincides with $J(f)$.

4.6.2   *From algebraic to ordinary Julia set*

In order to explain our expectation that algebraic Julia sets are related to conventional ones, we formulate two more hypotheses.

A map f defines two important subsets in **X**: the unions $\mathcal{O}_+(f)$ and $\mathcal{O}_-(f)$ of all stable and unstable periodic orbits of all orders – both are countable sets of points in **X**.

Hypothesis: For almost all $f \in \mathcal{M}$ the f-orbit of almost any point $x \in \mathbf{X}$ approaches $\mathcal{O}_+(f)$. Moreover, it approaches exactly one concrete stable orbit of particular order $n(x)$, which is a characteristic of the point x.

The pre-orbit tree of almost any point $x \in \mathbf{X}$ has infinitely many branches. Going backwards along particular branch we approach its inverse limit, or *origin*.

Hypothesis: For almost all $f \in \mathcal{M}$ and $x \in \mathbf{X}$ the origins of almost all *periodic* branches of grand f-orbit of x belong to $\mathcal{O}_-(f)$. Moreover, different periodic branches usually originate at different unstable orbits from $\mathcal{O}_-(f)$ and almost every orbit in $\mathcal{O}_-(f)$ is an origin of some branches of f-pre-orbit of x.

In other words, almost each grand orbit originates at the closure of *entire* $\mathcal{O}_-(f)$ and terminates at (tends to) a particular orbit in $\mathcal{O}_+(f)$. If these hypotheses are true, one and the same set $\mathcal{O}_-(f)$ is an origin of almost all grand orbits (not obligatory bounded). It is this set (or its closure, to be precise) that we call the boundary of the algebraic Julia set, $\partial J_A(f) = \overline{\mathcal{O}_-(f)}$. Periodic branches and unstable periodic orbits play the same role as periodic sequences (rational numbers) in the space of all sequences (real numbers), see Sec. 5.5 for more details.

Bounded grand orbits reach (not just tend to) a periodic orbit, however, in this case it can belong to $\mathcal{O}_-(f)$, not obligatory to $\mathcal{O}_+(f)$: bounded grand orbits are not in generic position in what concerns the "future". Still, they are not distinguished from the point of view of the "past" and can be used to study $J_A(f)$. Bounded grand orbits are convenient to deal with, because they consist of roots of $F_{n,s}$ (with all s) and can be studied in pure algebraic terms.

4.6.3 *Bifurcations of Julia set*

If we deform f (change the coefficients of the corresponding series) then the set of grand orbits move in $\mathbf{F}$ and may undergo the following two structural changes:

(i) merging of two distinct periodic orbits O', O'', which happens as soon as any pair of elements from these orbits merge,

(ii) merging of elements of the same pre-orbit x^-.

This merging of components of grand orbits results into splitting of $J_A(f)$ into disjoint components.

Case (i) corresponds to merging of the roots of F_n and ii) corresponds to merging of the roots of $F_{n,s}$. This means that f hits the discriminants $\mathcal{D}_{\text{on}} = \mathcal{D}(F_n)$ and $\mathcal{D}(F_{n,s})$ with $s \geq 1$ respectively, i.e. that some functions $F_n(x; f)$ or $F_{n,s}(x; f)$ become degenerate. If x is a multiple root of F_n then $F_n(x) = 0$ and $F_n'(x) = (f^{\circ n}(x) - x)' = 0$ so $(f^{\circ n}(x))' = 1$. Since n-periodic x is a stationary point for the map $f^{\circ n}$, then the last equation means that the points of the merging orbits O_n' and O_n'' change their stability type. Thus in the case when we can speak about convergence in $\mathbf{X}$, the language of "stability" is related (equivalent?) to that of "discriminants", but we will stick to the latter since it allows one to handle also the case ii) and the case of the arbitrary field $\mathbf{X}$ as well (which does not need to be full and have continuous and differentiable functions defined).

In the case (ii) some point of the grand orbit has degenerate preimage (two or more preimages coincide). A point $z_f \in \mathbf{X}$ has degenerate f-preimage when discriminant

$$D(f(x) - z_f) = 0. \tag{4.6}$$

This equation defines an important (multi-valued) map

$$z : \mathcal{M} \longrightarrow \mathbf{X},$$

associating a set of points $\{z_f\}$ – solutions to (4.6) – with every map $f \in \mathcal{M}$. Bifurcations of the type (ii) occur whenever some bounded grand orbit crosses this set, i.e. when

$$F_{n,s}(z_f; f) = 0$$

for some n and k.

The fact that all bifurcations of bounded grand orbits are either of the type (i) or of the type (ii), implies that discriminants $D(F_{n,s})$ are products of discriminants $D(F_n)$ and the new canonical functions $F_{n,r}(z)$ on $\mathcal{M}$, with $r \leq s$. Similarly, the resultants $R(F_{n,s}, F_{k,r})$ are made from $R(F_n, F_k)$ and various $F_{m,t}(z)$ with $m|n, k$ and $t \leq s, r$. Actual expressions are somewhat more involved, because the functions $F_{n,s}(z)$ are highly reducible (see Sec. 4.7 below) and their different components enter differently into formulas for particular resultants.

For description of these reductions it is important that the map z_f is intimately related to the critical points w_f of f. Indeed, Eq. (4.6) implies that $f(x) - z_f$ and $f'(x)$ have a common zero. The zeroes of derivative $f'(x)$ are critical points w_f:

$$f'(w_f) = 0,$$

and therefore

$$z_f = f(w_f),$$

i.e. there is a one-to-one correspondence between the points z_f and w_f. Moreover, from the definition of $F_{n,s}$ it follows that

$$F_{n,s}(z_f) = F_{n,s+1}(w_f). \tag{4.7}$$

4.7   On discriminant analysis for grand orbits

4.7.1   *Decomposition formula for $F_{n,s}(x; f)$*

Each zero of $F_{k,r}(x)$ satisfies $f^{\circ(k+r)}(x) = f^{\circ r}(x)$. Applying $f^{\circ k}$ to both sides of this equation, we obtain $f^{\circ(2k+r)}(x) = f^{\circ(k+r)}(x) = f^{\circ r}(x)$. Repeating this procedure several times we get $f^{\circ(mk+r)}(x) = f^{\circ r}(x)$ and, finally, applying $f^{\circ(s-r)}$ we obtain $f^{\circ(mk+s)}(x) = f^{\circ s}(x)$ for any $s \geq r$ and any m. This means that $F_{k,r}(x)$ is a divisor of any $F_{n,s}$, provided $k|n$ and $s \geq r$. Consequently, similarly to (4.1) we have the following decomposition of $F_{n,s}(x)$ into irreducible (for generic field $\mathbf{X}$ and map f) components:

$$F_{n,s}(x; f) = \prod_{k|n}^{\tau(n)} \prod_{r=0}^{s} G_{k,r}(x; f). \tag{4.8}$$

The irreducible function $G_{k,0}(x; f) = G_k(x; f)$ appeared already in (4.1). In particular, (4.8) states that

$$F_{1,1} = G_1 G_{1,1},$$
$$F_{1,2} = G_1 G_{1,1} G_{1,2},$$

$$\cdots$$

$$F_{1,s} = G_1 G_{1,1} G_{1,2} \ldots G_{1,s},$$

$$\cdots$$

$$F_{2,s} = (G_1 G_{1,1} G_{1,2} \ldots G_{1,s})(G_2 G_{2,1} G_{2,2} \ldots G_{2,s}),$$
$$F_{3,s} = (G_1 G_{1,1} G_{1,2} \ldots G_{1,s})(G_3 G_{3,1} G_{3,2} \ldots G_{3,s}),$$
$$F_{4,s} = (G_1 G_{1,1} G_{1,2} \ldots G_{1,s})(G_2 G_{2,1} G_{2,2} \ldots G_{2,s})(G_4 G_{4,1} G_{4,2} \ldots G_{4,s}),$$

$$\cdots$$

4.7.2 *Irreducible constituents of discriminants and resultants*

A direct analogue of (4.2) expresses discriminants $D(F_{n,s})$ through $D(G_{k,r})$ and the resultants $R(G_{k,r}, G_{k',r'})$. A less trivial thing is expressing these quantities through elementary constituents, which are, as already predicted, the familiar from Eq. (4.3) irreducible discriminants and resultants $d_k = d(G_k)$ and $r_{nk} = r(G_n, G_k)$, as well as the new quantities, which are irreducible components $w_{k,r}(f)$ of

$$W_{n,s}(f) = \prod_{w_f} G_{n,s}(w_f; f) \tag{4.9}$$

– the products of all values of $G_{r,s}(x; f)$ at all critical points w_f of f, $f'(w_f) = 0$.[2] Note that G, not F, enters the definition of W in (4.9), still after substitution of peculiar values of $x = w_f$ this quantity *often* gets

---

[2]We hope that the use of the same letter w for critical points $w_f \in \mathbf{X}$ and irreducible components $w_{k,r}(f) \in \mathcal{M}$ will not cause too much confusion.

further reducible, the first few reductions are:

$$W_n(f) = w_n(f) \quad \text{(irreducible)},$$

$$W_{1,s}(f) = w_1(f)w_{1,s}(f),$$

$$\begin{cases} W_{2,2r}(f) = w_{2,2r}(f) \quad \text{(irreducible)}, \\ W_{2,2r+1}(f) = w_2(f)w_{2,2r+1}(f), \end{cases}$$

$$\begin{cases} W_{3,3r}(f) = w_{3,3r}(f) \quad \text{(irreducible)}, \\ W_{3,3r+1}(f) = w_3(f)w_{3,3r+1}(f), \\ W_{3,3r+2}(f) = w_{3,3r+2}(f) \quad \text{(irreducible)}, \end{cases}$$

$$\begin{cases} W_{4,4r}(f) = w_{4,4r}(f) \quad \text{(irreducible)}, \\ W_{4,4r+1}(f) = w_4(f)w_{4,4r+1}(f), \\ W_{4,4r+2}(f) = w_{4,4r+2}(f) \quad \text{(irreducible)}, \\ W_{4,4r+3}(f) = w_{4,4r+3}(f) \quad \text{(irreducible)}, \end{cases}$$
$$\dots$$

(we actually checked most of these statements only for $n + s \leq 6$). Presumably, in general

$$\begin{cases} W_{n,s} = w_{n,s} \quad \text{for } s \neq 1 \text{ mod } n \quad \text{(irreducible)}, \\ W_{n,nr+1} = w_n w_{n,nr+1}. \end{cases} \tag{4.10}$$

4.7.3   *Discriminant analysis at the level* $(n,s) = (1,1)$*: basic example*

Let us begin the proof of (4.10) from the simplest case of $W_{1,1}$. The key point is that

$$G_{1,1}(x) = f'(x) \text{ mod } G_1(x). \tag{4.11}$$

Then, according to (4.36), the residue $R(G_{1,1}, G_1)$ can be expressed as a product over all critical points w_f, i.e. the roots of $f'(x)$:

$$r_{1|1,1} = R(G_1, G_{1,1}) \sim R(G_1, f') \sim \prod_w G_1(w) = W_1(f) = w_1(f). \tag{4.12}$$

Equation (4.12) is the first result of grand orbit discriminant calculus. It relies upon the relation (4.11), actually, on the fact that $f'(x)$ is residual of $G_{1,1}(x)$ division by $G_1(x)$: then, though $G_{1,1}(x)$ is not divisible by $G_1(x)$

as a function of x, $W_1(f) = \prod_w G_1(w)$ divides $W_{1,1}(f) = \prod_w G_{1,1}(w)$ as a functional of f. In order to see that

$$\frac{G_{1,1}(x) - f'(x)}{G_1(x)} = \frac{\frac{f(f(x)) - f(x)}{f(x) - x} - f'(x)}{f(x) - x} \qquad (4.13)$$

is non-singular at all zeroes of $G_1(x) = F_1(x) = f(x) - x$ it is enough to consider an infinitesimal variation of such root, $x = x_1 + \chi$, $f(x_1) = x_1$, and expand all functions at x_1 up to the second order in χ:

$$G_1(x) = F_1(x) = f(x) - x = (f'(x_1) - 1)\chi + \frac{1}{2}f''(x_1)\chi^2 + \ldots$$

$$= (f'(x_1) - 1)\chi \left(1 + \chi \frac{f''(x_1)}{2(f'(x_1) - 1)} + O(\chi^2)\right),$$

$$F_{1,1}(x) = f'(x_1)(f'(x_1) - 1)\chi + \frac{1}{2}f''(x_1)(f'(x_1)^2 + f'(x_1) - 1)\chi^2 + O(\chi^3)$$

$$= (f'(x_1) - 1)\chi \left(f'(x_1) + \frac{f''(x_1)\chi}{2(f'(x_1) - 1)}(f'(x_1)^2 + f'(x_1) - 1) + O(\chi^2)\right),$$

$$G_{1,1}(x) = \frac{F_{1,1}(x)}{G_1(x)} = \left(f'(x_1) + \frac{f''(x_1)(f'(x_1)^2 + f'(x_1) - 1)\chi}{2(f'(x_1) - 1)} + O(\chi^2)\right)$$

$$\times \left(1 - \frac{f''(x_1)\chi}{2(f'(x_1) - 1)} + O(\chi^2)\right) = f'(x_1) + \frac{1}{2}\chi f''(x_1)(f'(x_1) + 1) + O(\chi^2),$$

so that (4.13) becomes

$$\frac{f'(x_1) + \frac{1}{2}\chi f''(x_1)(f'(x_1) + 1) - f'(x_1) - \chi f''(x_1) + O(\chi^2)}{(f'(x_1) - 1)\chi + O(\chi^2)} = f''(x_1) + O(\chi)$$

and is finite at $\chi = 0$. This proves divisibility of $W_{1,1}(f)$ and allows one to introduce its irreducible constituent $w_{1,1}(f)$:

$$W_1(f) = \prod_{w_f} G_1(w_f) = w_1(f),$$

$$W_{1,1}(f) = \prod_{w_f} G_{1,1}(w_f) = w_1(f)w_{1,1}(f).$$

It enters expression for discriminant $D(G_{1,1})$:

$$D(G_{1,1}) \sim d_1^{d-1} w_{1,1} \qquad (4.14)$$

i.e. is actually the irreducible part of this discriminant:

$$d_{1,1} \sim w_{1,1}. \qquad (4.15)$$

Power $d-1$ in which $d_1 = d(G_1)$ enters (4.14) depends on the degree d of the map $f(x)$ (i.e. $f(x)$ is assumed to be a polynomial of degree d).
  Equations (4.15) and (4.12),

$$r_{1|1,1} \sim w_1,$$
$$d_{1,1} \sim w_{1,1}$$

together with

$$R(G_{1,1}, G_n) = 1, \quad \text{for } n > 1$$

are the outcome of discriminant/resultant analysis in the sector of G_1 and $G_{1,1}$, responsible for fixed points and their first pre-images.

4.7.4 *Sector $(n, s) = (1, s)$*

For a fixed point $x_1 = f(x_1)$ we denote $f' = f'(x_1)$, $f'' = f''(x_1)$. Then for $x = x_1 + \chi$ we have:

$$F_n(x) = \chi(f'^n - 1)\left(1 + \frac{1}{2}\chi f'' \frac{f'^{(n-1)}}{f' - 1} + O(\chi^2)\right)$$

and

$$G_k(x) = g_k(f')\left(1 + \frac{1}{2}\chi f'' h_k(f') + O(\chi^2)\right),$$

where $g_k(\beta)$ are irreducible circular polynomials (which will appear again in Chap. 6) and $h_k(\beta)$ are more sophisticated (with the single exception of h_1 they are also polynomials). The first several polynomials are:

$$
\begin{aligned}
g_1(\beta) &= \beta - 1 & h_1(\beta) &= 1 \\
g_2(\beta) &= \beta + 1 & h_2(\beta) &= 1 \\
g_3(\beta) &= \beta^2 + \beta + 1 & h_3(\beta) &= \beta + 1 \\
g_4(\beta) &= \beta^2 + 1 & h_4(\beta) &= \beta(\beta + 1) \\
g_5(\beta) &= \beta^4 + \beta^3 + \beta^2 + \beta + 1 & h_5(\beta) &= \beta^3 + \beta^2 + \beta + 1 \\
g_6(\beta) &= \beta^2 - \beta + 1 & h_6(\beta) &= \beta^4 + \beta^3 + \beta^2 - 1
\end{aligned}
$$

$$\cdots$$

Making use of $F_{n,s} = F_{n+s} - F_s$ and of decomposition formula (4.8), it is straightforward to deduce:

$$G_{1,1}(x) = \frac{F_2 - F_1}{G_1} = f' + \frac{1}{2}\chi f''(f' + 1) + O(\chi^2),$$

$$\frac{G_{1,1}(x) - f'(x)}{G_1(x)} = \frac{1}{2}\frac{f' + 1 - 2}{f' - 1}f'' + O(\chi) = \frac{1}{2}f'' + O(\chi),$$

as we already know. Here and below $f'(x) = f' + \chi f'' + O(\chi^2)$. Further, recursively,

$$G_{1,s}(x) = \frac{F_{s+1} - F_S}{G_1 G_{1,1}\ldots G_{1,s-1}} = f' + \frac{1}{2}\chi f''(f' + 1)f'^{s-1} + O(\chi^2),$$

$$\frac{G_{1,s}(x) - f'(x)}{G_1(x)} = \frac{1}{2}\frac{f'^s + f'^{s-1} - 2}{f' - 1}f'' + O(\chi)$$

$$= \frac{1}{2}f''(f'^{s-1} + 2f'^{s-2} + \ldots + 2) + O(\chi),$$

i.e. $G_{1,s}(x) = f'(x) \bmod G_1(x)$ and, as generalization of (4.11) and (4.12), we obtain for all s

$$r_{1|1,s} = R(G_1, G_{1,s}) \sim R(G_1, f') \sim \prod_{w_f} G_1(w_f; f) = w_1.$$

4.7.5 Sector $(n, s) = (2, s)$

For consideration of the sector $(n, s) = (2, s)$ we need to consider the vicinity of another stable point x_2, which belongs to the orbit of order two, i.e. satisfies $f(f(x_2)) = x_2$, but $\tilde{x}_2 = f(x_2) \neq x_2$. Denote $f' = f'(x_2)$, $\tilde{f}' = f'(\tilde{x}_2) = f'(f(x_2))$, $f'' = f''(x_2)$, $\tilde{f}'' = f''(\tilde{x}_2) = f''(f(x_2))$. For $x = x_2 + \chi$ we have:

$$F_1(x) = G_1(x) = (\tilde{x}_2 - x_2) + \chi(f' - 1) + \frac{1}{2}\chi^2 f'' + O(\chi^3),$$

$$F_2(x) = G_1(x)G_2(x) = \chi\left(f'\tilde{f}' - 1\right) + \frac{1}{2}\chi^2\left(f''\tilde{f}' + \tilde{f}''f'^2\right) + O(\chi^3),$$

$$F_3(x) = G_1(x)G_3(x) = (\tilde{x}_2 - x_2)$$
$$+ \chi\left(f'^2\tilde{f}' - 1\right) + \frac{1}{2}\chi^2\left(f''f'\tilde{f}' + \tilde{f}''f'^3 + f''(f'\tilde{f}')^2\right) + O(\chi^3),$$

$$\ldots$$

Then

$$G_{2,1}(x) = \frac{F_{2,1}}{G_1 G_2 G_{1,1}} = \frac{F_1(F_3 - F_1)}{(F_2 - F_1)F_2}$$

$$= -\frac{(\tilde{x}_2 - x_2) + \chi(f' - 1) + \frac{1}{2}\chi^2 f'' + O(\chi^3)}{(\tilde{x}_2 - x_2) - \chi(\tilde{f}' - 1)f' - \frac{1}{2}\chi^2 \left(f''(\tilde{f}' - 1) + \tilde{f}'' f'^2\right) + O(\chi^3)}$$

$$\cdot \frac{\chi\left(f'\tilde{f}' - 1\right)f' + \frac{1}{2}\chi^2 \left(\tilde{f}'' f'^3 + f''\left((f'\tilde{f}')^2 + f'\tilde{f}' - 1\right)\right) + O(\chi^3)}{\chi\left(f'\tilde{f}' - 1\right) + \frac{1}{2}\chi^2 \left(f''\tilde{f}' + \tilde{f}'' f'^2\right) + O(\chi^3)}$$

$$= -\left(1 + \frac{\chi(f'\tilde{f}' - 1)}{\tilde{x}_2 - x_2} + O(\chi^2)\right)\left(f' + \frac{1}{2}\chi f''(f'\tilde{f}' + 1) + O(\chi^2)\right),$$

so that

$$G_{21}(x) + f'(x) = -\chi(f'\tilde{f}' - 1)\left(\frac{1}{2}f'' + \frac{f'}{\tilde{x}_2 - x_2}\right) + O(\chi^2)$$

and

$$\frac{G_{21}(x) + f'(x)}{G_2(x)} = -\left(\frac{1}{2}f'' + \frac{f'}{\tilde{x}_2 - x_2}\right) + O(\chi)$$

is finite at $\chi = 0$. Therefore

$$r_{2|2,1} = R(G_2, G_{2,1}) \sim -R(G_2, f') \sim \prod_{w_f} G_2(w_f) = W_2(f) = w_2(f).$$

4.7.6   *Summary*

In the same way one can consider other quantities, complete the derivation of (4.10) and deduce the formulas for resultants and discriminants. Like in (4.14), these decomposition formulas depend explicitly on degree d of the map $f(x)$. In what follows

$$\#s = (d - 1)d^{s-1}$$

denotes the number of s-level pre-images of a point on a periodic orbit, which do not belong to the orbit. Also, $w_n(f)$ enter all formulas multiplied by peculiar d-dependent factors:

$\tilde{w}_n(f)$	$d = 2$	$d = 3$
$\tilde{w}_1 = d^d w_1,$	$2^2 = 4,$	$3^3 = 27,$
$\tilde{w}_2 = d^{d(d-1)} w_2,$	$2^2 = 4,$	$3^6 = 729,$
$\tilde{w}_3 = d^{d(d^2-1)} w_3,$	$2^6 = 64,$	$3^{24},$
$\tilde{w}_4 = d^{d^2(d^2-1)} w_4,$	$2^{12} = 4096,$	3^{72}
$\tilde{w}_5 = d^{d(d^4-1)} w_5,$	$2^{30},$	3^{240} ?
$\tilde{w}_6 = d^{d(d^2-1)(d^3+d-1)} w_6,$	$2^{54},$	3^{7176} ?

$$\cdots$$

(the last two columns in this table contain the values of numerical factors for $d = 2$ and $d = 3$, question marks label the cases which were *not* verified by explicit MAPLE simulations). Obviously, these factors are made from degree $N_d(n)$ of the map $G_n(x)$, which was considered in Sec. 4.1:

$$\tilde{w}_n(f) = d^{N_d(n)} w_n(f). \tag{4.16}$$

Non-trivial (i.e. not identically unit) resultants are:

$$R(G_m, G_{n,s}) = \begin{cases} \tilde{w}_n & \text{if } m = n, \\ 1 & \text{if } m \neq n \end{cases}$$

and

$$R(G_{k,s}, G_{n,s'}) = \begin{cases} R(G_k, G_n) = r_{k,n}^{k\#s} & \text{if } s' = s, \ k < n \ (\text{actually, } k|n) \\ \tilde{w}_n^{\#s} & \text{if } s < s' \end{cases}$$

or, in the form of Table 4.1.

Similarly, for discriminants:

$$D(G_{n,s}) \sim D(G_n)^{\#s} w_{11}^{\#s/\#1} \prod_{r=2}^{s} W_{n,r}^{\#s/\#r}$$

$$= \left(d_n^n \prod_{\substack{k|n \\ k<n}} r_{k,n}^{n-k} \right)^{\#s} w_n^{\delta(n,s)} \prod_{r=1}^{s} w_{n,r}^{\#s/\#r}. \tag{4.17}$$

Exponents in this expression are $\#s = (d-1)d^{s-1}$, $\#s/\#r = d^{s-r}$, and $\delta(n,s) = d^{r-1}\frac{d^{nr'}-1}{d^n-1} = \frac{d^{s-1}-d^{r-1}}{d^n-1}$ for $s = nr' + r$, $r', r > 0$ (if such r' and r do not exist, i.e. $s \leq n$, $\delta(n,s) = 0$). The first few values of $\delta(n,s)$ are listed in Table 4.2 (stars substitute too lengthy expressions).

Table 4.1

ns	11	12	13	14	21	22	23	24	31	32	33	34
1	$\tilde{w}_1$	$\tilde{w}_1$	$\tilde{w}_1$	$\tilde{w}_1$	1	1	1	1	1	1	1	1
2	1	1	1	1	$\tilde{w}_2$	$\tilde{w}_2$	$\tilde{w}_2$	$\tilde{w}_2$	1	1	1	1
3	1	1	1	1	1	1	1	1	$\tilde{w}_3$	$\tilde{w}_3$	$\tilde{w}_3$	$\tilde{w}_3$
4	1	1	1	1	1	1	1	1	1	1	1	1
5	1	1	1	1	1	1	1	1	1	1	1	1
. . .												
11	-	$\tilde{w}_1^{\#1}$	$\tilde{w}_1^{\#1}$	$\tilde{w}_1^{\#1}$	$r_{12}^{\#1}$	1	1	1	$r_{13}^{\#1}$	1	1	1
12	$\tilde{w}_1^{\#1}$	-	$\tilde{w}_1^{\#2}$	$\tilde{w}_1^{\#2}$	1	$r_{12}^{\#2}$	1	1	1	$r_{13}^{\#2}$	1	1
13	$\tilde{w}_1^{\#1}$	$\tilde{w}_1^{\#2}$	-	$\tilde{w}_1^{\#3}$	1	1	$r_{12}^{\#3}$	1	1	1	$r_{13}^{\#3}$	1
14	$\tilde{w}_1^{\#1}$	$\tilde{w}_1^{\#2}$	$\tilde{w}_1^{\#3}$	-	1	1	1	$r_{12}^{\#4}$	1	1	1	$r_{13}^{\#4}$
15	$\tilde{w}_1^{\#1}$	$\tilde{w}_1^{\#2}$	$\tilde{w}_1^{\#3}$	$\tilde{w}_1^{\#4}$	1	1	1	1	1	1	1	1
16	$\tilde{w}_1^{\#1}$	$\tilde{w}_1^{\#2}$	$\tilde{w}_1^{\#3}$	$\tilde{w}_1^{\#4}$	1	1	1	1	1	1	1	1
. . .												
21	$r_{12}^{\#1}$	1	1	1	-	$\tilde{w}_2^{\#1}$	$\tilde{w}_2^{\#1}$	$\tilde{w}_2^{\#1}$	1	1	1	1
22	1	$r_{12}^{\#2}$	1	1	$\tilde{w}_2^{\#1}$	-	$\tilde{w}_2^{\#2}$	$\tilde{w}_2^{\#2}$	1	1	1	1
23	1	1	$r_{12}^{\#3}$	1	$\tilde{w}_2^{\#1}$	$\tilde{w}_2^{\#2}$	-	$\tilde{w}_2^{\#3}$	1	1	1	1
24	1	1	1	$r_{12}^{\#4}$	$\tilde{w}_2^{\#1}$	$\tilde{w}_2^{\#2}$	$\tilde{w}_2^{\#3}$	-	1	1	1	1
25	1	1	1	1	$\tilde{w}_2^{\#1}$	$\tilde{w}_2^{\#2}$	$\tilde{w}_2^{\#3}$	$\tilde{w}_2^{\#4}$	1	1	1	1
26	1	1	1	1	$\tilde{w}_2^{\#1}$	$\tilde{w}_2^{\#2}$	$\tilde{w}_2^{\#3}$	$\tilde{w}_2^{\#4}$	1	1	1	1
. . .												
31	$r_{13}^{\#1}$	1	1	1	1	1	1	1	-	$\tilde{w}_3^{\#1}$	$\tilde{w}_3^{\#1}$	$\tilde{w}_3^{\#1}$
32	1	$r_{13}^{\#2}$	1	1	1	1	1	1	$\tilde{w}_3^{\#1}$	-	$\tilde{w}_3^{\#2}$	$\tilde{w}_3^{\#2}$
33	1	1	$r_{13}^{\#3}$	1	1	1	1	1	$\tilde{w}_3^{\#1}$	$\tilde{w}_3^{\#2}$	-	$\tilde{w}_3^{\#3}$
34	1	1	1	$r_{13}^{\#4}$	1	1	1	1	$\tilde{w}_3^{\#1}$	$\tilde{w}_3^{\#2}$	$\tilde{w}_3^{\#3}$	-
35	1	1	1	1	1	1	1	1	$\tilde{w}_3^{\#1}$	$\tilde{w}_3^{\#2}$	$\tilde{w}_3^{\#3}$	$\tilde{w}_3^{\#4}$
. . .												
41	$r_{14}^{\#1}$	1	1	1	$r_{24}^{2\,\#1}$	1	1	1	1	1	1	1
42	1	$r_{14}^{\#2}$	1	1	1	$r_{24}^{2\,\#2}$	1	1	1	1	1	1
43	1	1	$r_{14}^{\#3}$	1	1	1	$r_{24}^{2\,\#3}$	1	1	1	1	1
44	1	1	1	$r_{14}^{\#3}$	1	1	1	$r_{24}^{2\,\#4}$	1	1	1	1
45	1	1	1	1	1	1	1	1	1	1	1	1
. . .												
51	$r_{15}^{\#1}$	1	1	1	1	1	1	1	1	1	1	1
52	1	$r_{15}^{\#2}$	1	1	1	1	1	1	1	1	1	1
53	1	1	$r_{15}^{\#3}$	1	1	1	1	1	1	1	1	1
54	1	1	1	$r_{15}^{\#4}$	1	1	1	1	1	1	1	1
55	1	1	1	1	1	1	1	1	1	1	1	1
. . .												
61	$r_{16}^{\#1}$	1	1	1	$r_{26}^{2\,\#1}$	1	1	1	$r_{36}^{3\,\#1}$	1	1	1
62	1	$r_{16}^{\#2}$	1	1	1	$r_{26}^{2\,\#2}$	1	1	1	$r_{36}^{3\,\#2}$	1	1
63	1	1	$r_{16}^{\#3}$	1	1	1	$r_{26}^{2\,\#3}$	1	1	1	$r_{36}^{3\,\#3}$	1
64	1	1	1	$r_{16}^{\#4}$	1	1	1	$r_{26}^{2\,\#4}$	1	1	1	$r_{36}^{3\,\#4}$
65	1	1	1	1	1	1	1	1	1	1	1	1
. . .												

Table 4.2   The first few values of $\delta(n,s)$

$n \backslash s$	1	2	3	4	5	6	7	8	9	10	$\cdots$
1	0	1	$d+1$	d^2+d+1	d^3+d^2+d+1	$*$	$*$	$*$	$*$	$*$	
2	0	0	1	d	d^2+1	d^3+d	d^4+d^2+1	d^5+d^3+d	$*$	$*$	
3	0	0	0	1	d	d^2	d^3+1	d^4+d	d^5+d^2	d^6+d^3+1	
4	0	0	0	0	1	d	d^2	d^3	d^4+1	d^5+d	
5	0	0	0	0	0	1	d	d^2	d^3	d^4	
6	0	0	0	0	0	0	1	d	d^2	d^3	
7	0	0	0	0	0	0	0	1	d	d^2	
8	0	0	0	0	0	0	0	0	1	d	
9	0	0	0	0	0	0	0	0	0	1	
10	0	0	0	0	0	0	0	0	0	0	
$\cdots$											

The first few discriminants are listed in the subsequent table ($D_n = D(G_n)$, not to be confused with irreducible d_n; the last two columns contain numerical factors for $d = 2$ and $d = 3^3$).

ns	$D(G_{ns})$		$d=2$	$d=3$
11	$D_1^{d-1} w_{11}$	$= d_1^{d-1} w_{11}$	1	-3
12	$D_1^{d(d-1)} w_{11}^d W_{12}$	$= d_1^{d(d-1)} w_1 w_{11}^d w_{12}$	2^4	3^{27}
13	$D_1^{d^2(d-1)} w_{11}^{d^2} W_{12}^d W_{13}$	$= d_1^{d^2(d-1)} w_1 w_{11}^{d^2} w_{12}^d w_{13}$	2^{16}	3^{135}
14	$D_1^{d^3(d-1)} w_{11}^{d^3} W_{12}^{d^2} W_{13}^d W_{14}$	$= d_1^{d^3(d-1)} w_1^{d^2+d+1} w_{11}^{d^3} w_{12}^{d^2} w_{13}^d w_{14}$	2^{48}	3^{567}
15	$D_1^{d^4(d-1)} w_{11}^{d^4} W_{12}^{d^3} W_{13}^{d^2} W_{14}^d W_{15}$	$= d_1^{d^4(d-1)} w_1^{d^3+d^2+d+1} w_{11}^{d^4} w_{12}^{d^3} w_{13}^{d^2} w_{14}^d w_{15}$	2^{128}	?
21	$D_2^{d-1} W_{21}$	$= d_2^{2(d-1)} r_{12}^{d-1} w_{21}$	1	3^6
22	$D_2^{d(d-1)} W_{21}^d W_{22}$	$= d_2^{2d(d-1)} r_{12}^{d(d-1)} w_{21}^d w_{22}$	2^4	3^{54}
23	$D_2^{d^2(d-1)} W_{21}^{d^2} W_{22}^d W_{23}$	$= d_2^{2d^2(d-1)} r_{12}^{d^2(d-1)} w_2 w_{21}^{d^2} w_{22}^d w_{23}$	2^{16}	?
24	$D_2^{d^3(d-1)} W_{21}^{d^3} W_{22}^{d^2} W_{23}^d W_{24}$	$= d_2^{2d^3(d-1)} r_{12}^{d^3(d-1)} w_2^d w_{21}^{d^3} w_{22}^{d^2} w_{23}^d w_{24}$	2^{48}	?
25	$D_2^{d^4(d-1)} W_{21}^{d^4} W_{22}^{d^3} W_{23}^{d^2} W_{24}^d W_{25}$	$= d_2^{2d^4(d-1)} r_{12}^{d^4(d-1)} w_2^{d^2+1} w_{21}^{d^4} w_{22}^{d^3} w_{23}^{d^2} w_{24}^d w_{25}$	2^{128}	?
31	$D_3^{d-1} W_{31}$	$= d_3^{3(d-1)} r_{13}^{2(d-1)} w_{31}$	1	3^{24}
32	$D_3^{d(d-1)} W_{31}^d W_{32}$	$= d_3^{3d(d-1)} r_{13}^{2d(d-1)} w_{31}^d w_{32}$	2^{12}	?
33	$D_3^{d^2(d-1)} W_{31}^{d^2} W_{32}^d W_{33}$	$= d_3^{3d^2(d-1)} r_{13}^{2d^2(d-1)} w_{31}^{d^2} w_{32}^d w_{33}$	2^{48}	?
34	$D_3^{d^3(d-1)} W_{31}^{d^3} W_{32}^{d^2} W_{33}^d W_{34}$	$= d_3^{3d^3(d-1)} r_{13}^{2d^3(d-1)} w_3 w_{31}^{d^3} w_{32}^{d^2} w_{33}^d w_{34}$	2^{144}	?
35	$D_3^{d^4(d-1)} W_{31}^{d^4} W_{32}^{d^3} W_{33}^{d^2} W_{34}^d W_{35}$	$= d_3^{3d^4(d-1)} r_{13}^{2d^4(d-1)} w_3^d w_{31}^{d^4} w_{32}^{d^3} w_{33}^{d^2} w_{34}^d w_{35}$	2^{384}	?
41	$D_4^{d-1} W_{41}$	$= d_4^{4(d-1)} r_{14}^{3(d-1)} r_{24}^{2(d-1)} w_{41}$	1	?
42	$D_4^{d(d-1)} W_{41}^d W_{42}$	$= d_4^{4d(d-1)} r_{14}^{3d(d-1)} r_{24}^{2d(d-1)} w_{41}^d w_{42}$	2^{24}	?
43	$D_4^{d^2(d-1)} W_{41}^{d^2} W_{42}^d W_{43}$	$= d_4^{4d^2(d-1)} r_{14}^{3d^2(d-1)} r_{24}^{2d^2(d-1)} w_{41}^{d^2} w_{42}^d w_{43}$	2^{96}	?
44	$D_4^{d^3(d-1)} W_{41}^{d^3} W_{42}^{d^2} W_{43}^d W_{44}$	$= d_4^{4d^3(d-1)} r_{14}^{3d^3(d-1)} r_{24}^{2d^3(d-1)} w_{41}^{d^3} w_{42}^{d^2} w_{43}^d w_{44}$	?	?
45	$D_4^{d^4(d-1)} W_{41}^{d^4} W_{42}^{d^3} W_{43}^{d^2} W_{44}^d W_{45}$	$= d_4^{4d^4(d-1)} r_{14}^{3d^4(d-1)} r_{24}^{2d^4(d-1)} w_4 w_{41}^{d^4} w_{42}^{d^3} w_{43}^{d^2} w_{44}^d w_{45}$	?	?
51	$D_5^{d-1} W_{51}$	$= d_5^{5(d-1)} r_{15}^{4(d-1)} w_{51}$	1	?
52	$D_5^{d(d-1)} W_{51}^d W_{52}$	$= d_5^{5d(d-1)} r_{15}^{4d(d-1)} w_{51}^d w_{52}$	2^{60}	?
61	$D_6^{d-1} W_{61}$	$= d_6^{6(d-1)} r_{16}^{5(d-1)} r_{26}^{4(d-1)} r_{36}^{3(d-1)} w_{61}$	1	?
62	$D_6^{d(d-1)} W_{61}^d W_{62}$	$= d_6^{6d(d-1)} r_{16}^{5d(d-1)} r_{26}^{4d(d-1)} r_{36}^{3d(d-1)} w_{21}^d w_{22}$	?	?
. . .				

4.7.7   On interpretation of $w_{n,k}$

Intersections with the set $\{z_f\} = f(\{w_f\})$, which is f-image of the set of critical points of f are responsible for degenerations of pre-orbits. Let us describe what happens when $\{z_f\}$ is crossed by pre-orbits of different levels.

- $\{G_n(f) = 0\} \bigcap \{z_f\} \neq \emptyset.$

---

[3]Numerical factors are equal to $d^{d^s(sd-s-1)\#_n(d)}$, where the sequences $\#_n(d)$ are $1, 1, 3, 6, 15, \ldots$ for $d = 2$ and $1, 2, \ldots$ for $d = 3$. One can observe that $\#_n(d) = \deg_c G_n(x; x^d + c)$, but the reason for this, as well as the very origin of numerical factors, here and in Eq. (4.16), remain obscure.

Let A be the intersection point, denote its pre-images along the periodic orbit through $A^{-1}, A^{-2}, \ldots, A^{-n} = A$, and the pre-images of level s on the pre-image tree through $B^{-s}_{i_1 \ldots i_{\#s}}(A)$, where indices i can be further ordered according to the tree structure. Points on the level s of the tree, rooted at pre-image A^{-l} will be denoted through $B^{-s}_{i_1 \ldots i_{\#s}}(A^{-l})$. If $A \in \{z_f\}$, then two of its pre-images coincide. There are two possibilities: either these coinciding preimages belong to pre-orbit tree, say $B^{-1}_1(A) = B^{-1}_2(A)$, or one of them lie on the orbit, say $B^{-1}_1(A) = A^{-1}$. In the first case $w_{n,1}(f) = 0$, in the second case $w_n(f) = 0$, and the fact that only two such possibilities exist is reflected in the decomposition formula

$$\prod_{z_f} G_n(z_f) = \prod_{w_f} G_{n,1}(w_f) = W_{n,1}(f) = w_n(f) w_{n,1}(f)$$

(products on the l.h.s. are needed to build up quantities, which depend on the coefficients of $f(x)$, not on their irrational combinations, entering expressions for individual points z_f and w_f – the latter are roots of $f'(x)$). Let us analyze these two cases in more detail.

○ $w_n(f) = 0$. See Fig. 4.1.
Since $B^{-1}_1(A) = A^{-1}$ exactly one point from $\{G_{n,1} = 0\}$ coincides with exactly one point on $\{G_n = 0\}$, thus the resultant has simple zero,

$$R(G_n, G_{n,1}) = r_{n|n,1} = \tilde{w}_n \sim w_n.$$

Numeric d-dependent factor, distinguishing between $\tilde{w}_n$ and w_n requires separate explanation.

Pre-images of $B^{-1}_1(A)$ – the set of d (different!) points $B^{-2}_1(A), \ldots, B^{-2}_d(A)$ from $\{G_{n,2} = 0\}$ should coincide pairwise with pre-images of A^{-1}, which are $d - 1$ points $B^{-1}_1(A^{-1}), \ldots, B^{-1}_{d-1}(A^{-1})$ from $\{G_{n,1} = 0\}$ and the point A^{-2} on the orbit, i.e. from $\{G_n = 0\}$. Thus the corresponding resultants will have zeroes of orders $\#1 = d - 1$ and one respectively:

$$R(G_{n,1}, G_{n,2}) = r^{\#1}_{n,1|n,2} \sim w^{\#1}_n$$

and

$$R(G_n, G_{n,2}) = r_{n|n,2} \sim w_n.$$

Next pre-images of A^{-1} consist of $A^{-3} \in \{G_n = 0\}$, $\#1 = d - 1$ points $B^{-1}(A^{-2}) \in \{G_{n,1} = 0\}$ and $\#2 = d(d-1)$ points $B^{-1}(A^{-2}) \in \{G_{n,2} = 0\}$, while those of B^{-1}_1 are $\#3/\#1 = d^2$ points from $\{B^{-3}(A)\}$. Each of d^2 points from the A^{-1} second pre-image should coincide with one of the d^2 points from that of B^{-1}_1. This provides relations

$$R(G_{n,2}, G_{n,3}) = r^{\#2}_{n,2|n,3} \sim w^{\#2}_n,$$

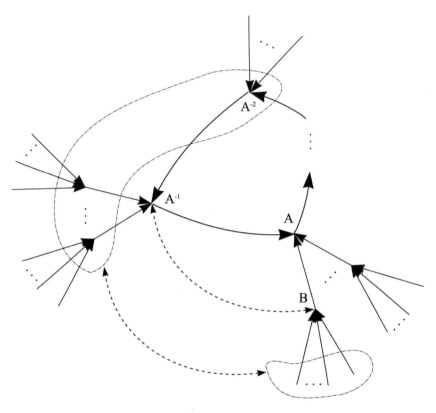

Fig. 4.1 A periodic orbit of order n and grand-orbit trees, rooted at the points of the orbit. $w_n(f)$ vanishes when (i) a point A on the orbit coincides with an "irreducible point" from the set $\{z_f\}$, which is f-image of the set of critical points $\{w_f\}$, and (ii) when one of the coinciding first pre-images of A is also a point on the orbit. Arrows show the pre-image points of different levels, which coincide when $A \in \{z_f\}$ and $w_n(f) = 0$, i.e. when $A^{-1} = B$.

$$R(G_{n,1}, G_{n,3}) = r_{n,1|n,3}^{\#1} \sim w_n^{\#1}$$

and

$$R(G_n, G_{n,3}) = r_{n|n,3} \sim w_n.$$

Continuing along the same line, we deduce:

$$R(G_{n,r}, G_{n,s}) = r_{n,r|n,s}^{\#r} \sim w_n^{\#r}, \quad \text{for } r < s$$

and

$$R(G_n, G_{n,s}) = r_{n|n,s} \sim w_n.$$

Actually, this exhausts the set of non-trivial (non-unit) resultants for orbits of coincident periods n, but does not exhaust the possible appearances of w_n: it will show up again in discriminants, see below.

○ $w_{n,1}(f) = 0$. See Fig. 4.2.

In this case we deal with a single tree, rooted at $A \in G_n \cap \{z_f\}$ and thus we can safely omit reference to A in $B(A)$. Thus, the starting point is $B_1^{-1} = B_2^{-1}$. (In particular, there is nothing to discuss in the case of $d = 2$, when $\#1 = 1$, there are no *a priori* different points at the level $s = 1$, which could occasionally coincide, and all $w_{n,1} = 1$.) We can conclude, that discriminant $D(G_{n,1}) \sim w_{n,1}$.

This is not the full description of discriminant, because it also contains contributions coming from intersections of different periodic orbits, which cause the corresponding points on pre-orbit trees to coincide as well. These factors were already evaluated in Sec. 4.3 above, in application to discriminant of $G_{n,s}$ they should just be raised to the power $\#s$, counting the number of pairs of points which are forced to coincide at level s when their roots on the orbits merge. Thus finally

$$D(G_{n,1}) \sim D(G_n)^{\#1} w_{n,1} = w_{n,1} \left(d_n^n \prod_{\substack{k|n \\ k<n}} r_{k,n}^{n-k} \right)^{\#1},$$

and the remaining undetermined factor is just a constant on $\mathcal{M}$, which depends on d, but not on f. The same argument explains the expressions for non-trivial resultants

$$R(G_{n,s}, G_{k,s}) = R(G_n, G_k)^{\#s} = r_{n,k}^{k\#s}, \quad k|n, \ k < n.$$

Considering next pre-images of B_1^{-1} and B_2^{-1}, we obtain at level s the two coincident sets of $\#s/\#1 = d^{s-1}$ points each and thus $D(G_{n,s}) \sim w_{n,1}^{d^{s-1}} D(G_n)^{\#s}$. However, many more factors enter in the expression for higher discriminants: all $w_{n,r}$ with $r < s$ contribute, and, remarkably, w_n also contributes when $s > n$ – in accordance with decomposition formula (4.10). To see all this we should return to the very beginning and consider the intersections of *pre-orbits* with the set $\{z_f\}$.

● $\{G_{n,r-1}(f) = 0\} \bigcap \{z_f\} \neq \emptyset$, $r > 1$. See Fig. 4.3.

Let $B_1^{-(r-1)} \in \{z_f\}$. This means that some two preimages of this point coincide, say $B_1^{-r} = B_2^{-r}$, and at level $s \geq r$ there will be two coinciding sets, of $\#s/\#r = d^{s-r}$ points each. Since $G_{n,r-1}(z_f; f) =$

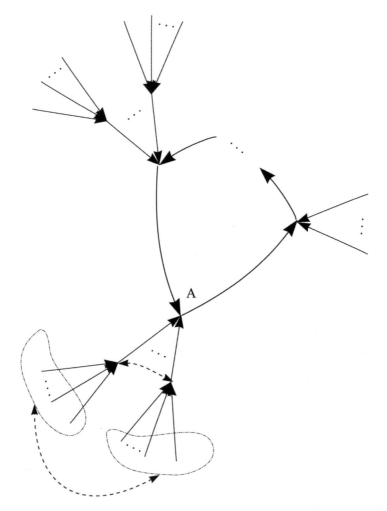

Fig. 4.2 The same periodic orbit of order n and the same point A, intersecting $\{z_f\}$ as in the previous Fig. 4.1. $w_{n,1}(f)$ vanishes when (i) a point A coincides with a point from $\{z_f\}$, and (ii) when both coinciding first pre-images of A belong to pre-image tree, rooted at A. Arrows show the pre-image points of different levels, which coincide when $A \in \{z_f\}$ and $w_{n,1}(f) = 0$.

$G_{n,r-1}(f(w_f); f) \sim G_{n,r}(w_f; f)$, all this happens when $W_{n,r}(f) = 0$, and we conclude that $D(G_{n,s}) \sim W_{n,r}^{d^{s-r}}$ for any $r \leq s$. We can further use (4.10) to substitute $W_{n,r}(f)$ by $w_{n,r}(f)$ for $r \neq 1 \bmod n$ and by $w_{n,r}(f)w_n(f)$ for

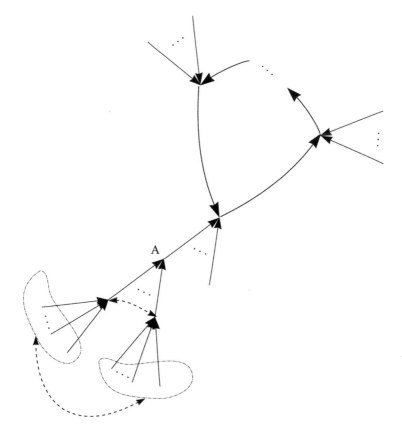

Fig. 4.3 A point which intersects $\{z_f\} = f(\{w_f\}$ belongs to the $r - 1$-th level of the grand orbit tree, $r > 1$. Arrows show the pre-image points of different levels, which coincide in this situation.

$r = 1 \bmod n$. Collecting everything together we finally reproduce (4.17):

$$D(G_{n,s}) \sim D(G_n)^{\#s} w_{11}^{\#s/\#1} \prod_{r=2}^{s} W_{n,r}^{\#s/\#r}$$

$$= \left(d_n^n \prod_{\substack{k|n \\ k<n}} r_{k,n}^{n-k} \right)^{(d-1)d^{s-1}} w_n^{\delta(n,s)} \prod_{r=1}^{s} w_{n,r}^{d^{s-r}} .$$

Index $\delta(n, s)$ is obtained by summation over all $1 < r \leq s$, such that $r = 1 \bmod n$, of the weights $\#s/\#r = d^{s-r}$. Each term in this sum corresponds to the tree, warped $\frac{n}{r-1}$ times around the periodic orbit, see Fig. 4.4. For $s = nr' + r$ and $r', r > 0$ we have

$$\delta(n, s) = d^{r-1} \frac{d^{nr'} - 1}{d^n - 1} = \frac{d^{s-1} - d^{r-1}}{d^n - 1}.$$

If such r' and r do not exist, i.e. if $s \leq n$, $\delta(n, s) = 0$.

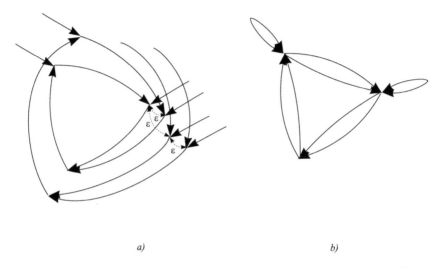

a)   b)

Fig. 4.4   a) Pre-orbit tree, warped around the periodic orbit of order $n = 3$. b) The limit degenerate orbit with multiplicities.

4.8   Combinatorics of discriminants and resultants

Systematic analysis of orbits of the map $f : \mathbf{X} \to \mathbf{X}$ includes determination of the following characteristics, used in our presentation in Chap. 3:

• Number $N_n(f)/n$ of periodic orbits of order n, which is n times smaller than the number $N_n(f)$ of roots of $G_n(x; f)$,

$$N_n(f) = \deg_x[G_n].$$

• Number $S(n; f)$ of different combinations of periodic orbits of order n which *can* be stable for some f. (It deserves emphasizing, that some orbits are never stable: for $f = x^2 + c$ only one of the two fixed points – orbits of order one – has non-vanishing stability domain in the plane of complex c.)

• Number $\tilde{\mathcal{N}}_n(f)$ of elementary components σ in stability domain S_n. It is also useful to count separately the numbers $\mathcal{N}_n^{(p)}(f)$ of elementary components σ at level p,

$$\tilde{\mathcal{N}}_n(f) = \sum_{p=0}^{\infty} \mathcal{N}_n^{(p)}(f).$$

• Number $\mathcal{N}_n(f)$ of solutions to the equation $G_n(x=0, f) = 0$, considered as an equation for f. Solutions of this equation for particular families of maps $\mu \subset \mathcal{M}$ are also important, because they define some points inside the σ-domains (and thus – approximately – the location of domains). If the family μ is one-parametric and parameter is c, then

$$\mathcal{N}_n(f) = \deg_c[G_n].$$

Presumably, $\tilde{S}(n; f) = \tilde{\mathcal{N}}_n(f) = \mathcal{N}_n(f)$.

• Number of zeroes of reduced discriminant $d_n = d(G_n)$, introduced in Sec. 4.3. For one-parametric families of maps f the number of zeroes is equal to $\deg_c[d_n]$. Zeroes of d_n are associated with self-intersections of periodic orbits and define the cusps on various components. Usually the number q of cusps depends only on the level of p of the component $\sigma_n^{(p)}$ (and not on n and other parameters like m_i and α_i).

• Number of zeroes of reduced resultants $r_{n,k} = r(G_n, G_k)$. For one-parametric families of maps f the number of zeroes is equal to $\deg_c[r_{n,k}]$. Each zero describes some merging of the elementary σ-domains. The number of zeroes is related to the number of α-parameters.

All these numbers are related by sum rules:

$$\sum_{p=0}^{\infty} \mathcal{N}_n^{(p)} = \deg_c[G_n(w_c, c)],$$

$$\sum_{p=0}^{\infty} q(p)\mathcal{N}_n^{(p)} = \deg_c[d_n],$$

$$\sum_{p=1}^{\infty} \mathcal{N}_n^{(p)} = \sum_{k|n} \deg_c[r_{n,k}]. \tag{4.18}$$

Example. For f which is a polynomial of degree d, $f(x) = x^d + c$, we have:

• $n = 1$:

\# of orbits: d,

\# of "potentially-stable" orbits: $S(1; x^d + c) = 1$.

Since $G_1(0; f) = f(0) = a_0 = c$, the number of solutions to $G_1(0; c) = 0$ is 1.

Discriminant $d(G_1) = D(G_1) = D(F_1)$ vanishes when simultaneously $x^d + c = x$ and $dx^{d-1} = 1$: there are $d - 1$ solutions, thus discriminant has $d - 1$ zeroes.

- $n = 2$:

\# of orbits: $\frac{d^2 - d}{2}$,

\# of "potentially-stable" orbits: $S(2; x^d + c) = d - 1$.

Since $G_2(0; c) = c^{d-1} + 1 = 0$, the number of solutions to $G_2(0; c) = 0$ is 1.

Discriminant $d(G_2) = \sqrt{\frac{D(G_2)}{R(G_1, G_2)}}$ has $(d-1)(d-2)$ zeroes. Each of the $d - 1$ elementary stability domains for orbits of order two has $d - 2$ cusps.

We list much more data of this kind for particular families of maps in the tables in Chap. 6. Similar analysis can be performed for bounded grand orbits.

4.9   Shapes of Julia and Mandelbrot sets

4.9.1   *Generalities*

The shape of Mandelbrot can be investigated with the help of stability constraints (2.5):

$$\begin{cases} |F_n'(x; f) + 1| = 1, \\ G_n(x; f) = 0. \end{cases} \tag{4.19}$$

Exclusion of x from these equations provides a real-codimension-one subspace in $\mu \subset \mathcal{M}$, which defines the boundary ∂S_n of stability domain S_n. Its connected components are boundaries of the elementary domains σ. Enumerating all n we can obtain in this way the entire boundary of Mandelbrot space $M(\mu)$.

Zeroes of all discriminants $D(G_n)$ belong to this boundary. Indeed, if $D(G_n) = 0$ there is a point $x_n \in \mathbf{X}$ which is common root of G_n and G_n': $G_n(x_n) = G_n'(x_n) = 0$ (actually, there is the entire set of such points, labeled by additional α-index). Then, since $F_n = \prod_{k|n} G_k = G_n \tilde{F}_n$, we have $F_n'(x_n) = G_n(x_n)\tilde{F}_n'(x_n) + G_n'(x_n)\tilde{F}_n(x_n) = 0$ and thus both equations

(4.19) are satisfied. Furthermore, since according to (4.3), a zero of every resultant $R(G_n, G_{n/m})$ is also a zero of $D(G_n)$, we conclude that zeroes of all such resultants also belong to ∂M. In the world of iterated maps only resultants of this kind are non-trivial, $R(G_n, G_k) = const$ unless $k|n$ or $n|k$.

One possibility to define the shape of $M(\mu)$ is just to **plot enough zeroes of discriminants and resultants** and – since they are dense in $\partial M(\mu)$ – this provides approximation with any desired accuracy. Advantage of this approach is that it is pure algebraic and – once formulated – does not contain any reference to stability equations and to the notion of stability at all. The disadvantage is that – at least in presented form – it does not separate points from different elementary domains: zeroes of each particular resultant are distributed between many elementary domains, which can even belong to different disconnected components $M_{k\alpha}$ of M. Also, it does not describe directly the boundaries of elementary domains, which – outside a few cusps, located at zeroes of reduced discriminants $d(G_n)$ – are smooth curves of peculiar (multi-cusp) cardioid-like form. Finally, it does not explain the similarity of various components $M_{k\alpha}$, which – for reasonably chosen families μ – belong to a set of universality classes labeled by $\mathbf{Z_{d-1}}$-symmetries.

Another option is to use equations (4.19) more intensively and transform them into less transcendental form, making reasonable re-parameterizations and approximations. The choice of parameterizations, however, can impose restrictions on the families μ, but instead one can move much further in explicit description of constituents of Mandelbrot and even Julia sets.

In the case of Julia sets the situation with non-algebraic approaches is somewhat more difficult. These sets do not have elementary smooth constituents, except for exactly at the bifurcation points, i.e. at the boundary points of Mandelbrot set, but even on ∂M the decomposition of $J(f)$ depends discontinuously on the point at the boundary. As soon as one goes inside Mandelbrot set, infinitely many different smooth structures interfere and only some traces of them can be seen approximately. The structure of Julia sets is pure algebraic, not smooth (or, if one prefers, consistent with d-adic rather than with $\mathbf{C}$-topology). The boundary of Julia set for given f is formed by solutions of all the equations $F_{n,s} = 0$ for all n and s, with exclusion of a few orbits, which are stable for this f – and this remains the best existing *constructive* definition of Julia set in general situation. Also, approximate methods can be used to analyze some features of Julia sets.

4.9.2 *Exact statements about 1-parametric families of polynomials of power-d*

Assuming that $\mu \subset \mathcal{P}_d$ and $\dim_{\mathbf{C}}\mu = 1$, we can label the maps f in the family by a single parameter c: $\mu = \{f_c\}$. For given $c \in M_1(\mu)$ define:

$$F'_n(x;c) + 1 = e^{i\varphi(d-1)}. \tag{4.20}$$

This relation maps the roots of $G_n(x;c)$ (the periodic orbits of order n) into the unit circle, parameterized by the angle φ. Considering all possible n and taking the closure, one can extend this map to entire boundary of Julia set. For $c \in M_1$ this provides a one-to-one correspondence between ∂J_c and the unit circle. If $c \in M_{k\alpha}$, analogous expression,

$$F'_{nk}(x;c) + 1 = e^{i\varphi(d-1)},$$

relates a *part* of ∂J_c and unit circle. Other parts are mapped onto additional circles. Parameterization (4.20) is adjusted to describe the boundary of Mandelbrot set: it explicitly solves the first equation in (4.19) and the function $x(\varphi)$ can be substituted into $G_n(x;c) = 0$ to obtain $c(\varphi)$. The map $c(\varphi)$ has several branches, associated with different elementary domains σ from stability domain S_n. This program can be realized in certain approximation.

Before going to approximations, let us give an example of the exact statement. The map $S^1 \to \partial\sigma$ is singular (there are cusps on the boundary $\partial\sigma$) whenever

$$\begin{cases} \left|\frac{\partial c}{\partial \varphi}\right| = (d-1)\left|\frac{G'_n}{H_n}\right| = 0, \\ G_n = 0. \end{cases} \tag{4.21}$$

Here $\dot{G}_n = \partial G_n/\partial c$ and

$$H_n = \{F'_n, G_n\} = \dot{F}'_n G'_n - F''_n \dot{G}_n.$$

Solutions to this system are all zeroes of the resultant

$$R(G_n, G'_n) \sim D(G_n) = d^n(G_n)\prod_{k|n} r^{n-k}(G_n, G_k) \tag{4.22}$$

which are *not* simultaneously zeroes of another resultant, $R(G'_n, H_n)$, or, which is much simpler to check, $R(G_n, H_n)$. Actually roots of all reduced resultants $r(G_n, G_k)$ are excluded, and the cusps of $\partial\sigma$ are at zeroes of reduced discriminants $d(G_n)$. Indeed, whenever $r(G_n, G_k) = 0$ for $k|n$, there is a common zero $\tilde{x}$ of the two functions $G_n(\tilde{x}) = G_k(\tilde{x}) = 0$. Also, because of (4.22), $G'_n(\tilde{x}) = 0$, therefore $H_n(\tilde{x}) = -F''_n \dot{G}_n(\tilde{x})$. Both G_n and

G_k enter the product (4.1) for F_n, $F_n \sim G_n G_k$ and two derivatives in F_n'' are not enough to eliminate G_n, G_k and G_n' from the product, therefore also $F_n''(\tilde{x}) = 0$ and $H_n(\tilde{x}) = 0$.

For polynomial maps elementary domains of Mandelbrot set $M(c)$ belong to universality classes, which are represented by multi-cusp cardioids (cycloids), described by the function $\varepsilon(\varphi)$ in complex ε-plain

$$C_0: \quad \varepsilon = e^{i\varphi},$$

$$C_{d-1}: \quad \varepsilon = e^{i\varphi} - \frac{1}{d}e^{i\varphi d}, \quad \text{for } d > 1.$$

The real curve C_{d-1} has discrete symmetry

$$\mathbf{Z_{d-1}}: \quad \varphi \to \varphi + \theta, \quad \varepsilon \to \mathbf{e^{i\theta}}\varepsilon,$$

it is singular, $|\partial \varepsilon / \partial \varphi| = 0$, provided that $\varphi = \frac{2\pi i k}{d-1}$, $k = 0, 1, \ldots, d-2$, i.e. the number of cusps is $d - 1$.

Generically, the boundaries of elementary domains belong to the class C_1 for level $p = 0$ and to C_0 for all other levels $p \geq 1$. However, for special families, like $f_c = x^d + c$, the situation will be different: at level $p = 0$ the elementary domains $\sigma^{(0)}$ can belong to the class C_{d-1} (has $d - 1$ cusps), while at higher levels it decreases to C_{d-2} (has $d - 2$ cusps). If symmetry is broken by adding lower powers of x with small coefficients to f_c, the small-size components $M_{k\alpha}$ fall into generic $C_0 \oplus C_1$ class, and the bigger the symmetry-breaking coefficients, the bigger are components that switch from $C_{d-2} \oplus C_{d-1}$ to $C_0 \oplus C_1$ class.

4.9.3 *Small-size approximation*

In order to explain how cardioids arise in description of elementary domains, we continue with the example of one-parameter family of polynomials of degree d and describe an *approximation*, which can be used to explain the above mentioned results. It is based on expansion of equations (4.19) around the point $(x, c) = (w, c_n)$, where w_c is a critical point of $f_c(x)$ and c_n is the "center" of an elementary domain from S_n, defined from solution of the equations

$$\begin{cases} f_c'(w_c) = 0, \\ G_n(w_c; c) = 0. \end{cases} \tag{4.23}$$

We now substitute into (4.19) $x = w + \chi$ and $c = c_n + \varepsilon$, expand all the functions in powers of ε and χ, and leave the first relevant approximation. Typically $|\chi| < d^{-n}$ and $|\varepsilon| < d^{-n}$, actually there is much stronger damping for elementary domains belonging to "remote" $M_{k\alpha}$, thus approximation can be numerically very good in most situations.

One can even promote this approach to alternative description of Mandelbrot set as a blow up of the system of points (4.23): each point gets surrounded by its own domain σ. Some of these domains get large enough to touch each other – and form the connected components $M_{k\alpha}$,– some remain disconnected topologically, but get instead connected by *trails*. Consistency of such approach and the one, based on resultant's zeroes, requires that the number of solutions to (4.23) with given n coincides with the total number of zeroes of $r_{nm}(c)$ with all $m < n$ plus the number of components $M_{n\alpha}$ growing from the elementary domain $\sigma^{(0)}[n\alpha]$,

$$\deg_c G_n(w_c, c) = |\nu_n^{(0)}| + \sum_{m|n} \deg_c r_{n,m}(c) \quad \forall \text{ 1-parametric families } \mu$$

(4.24)

where $\#F(c) = \deg_c F(c)$ denotes the number of roots of $F(c)$, which for polynomial $F(c)$ is equal to its degree (for multi-parametric families this is the relation between the degrees of algebraic varieties). A similar sum rule relates the number of roots of reduced determinant $d_n(c)$ and the numbers $\mathcal{N}_n^{(p)}$ of solutions to (4.23) with $q(p)$ cusps:

$$\sum_p q(p)\mathcal{N}_n^{(p)} = \deg_c d_n(c).$$

(4.25)

4.9.4   *Comments on the case of* $f_c(x) = x^d + c$

In this case there is a single critical point $w_f = 0$.

Stability domain S_1 for orbits of order one consists of the single elementary domain, $S_1 = \sigma[1]$ and is defined by a system

$$d|x^{d-1}| = 1,$$
$$x^d - x + c = 0.$$

(4.26)

Its boundary can be parameterized as follows: $x = d^{-1/(d-1)} e^{i\phi}$, and

$$\partial S_1: \quad c = x - x^d = \frac{1}{d^{1/(d-1)}} \left(e^{i\phi} - \frac{1}{d} e^{i\phi d} \right).$$

(4.27)

This is a curve with a cusp at $c = d - 1/d^{d/(d-1)}$ and the symmetry group $\mathbf{Z_{d-1}}$, the shift $\phi \to \phi + \frac{2\pi}{d-1}$ multiplies c by $e^{\frac{2\pi i}{d-1}}$ (so that there are actually $d - 1$ cusps). Julia sets have another symmetry, $\mathbf{Z_d}$.

We use this example to show how approximate similarity of different components $M_{k\alpha}$ of Mandelbrot set can be explained by "perturbation theory" from the previous subsection 4.9.3. We discuss just a very particular part of the story: the central domains $\sigma^{(0)}[k\alpha]$, obtained by the procedure from Sec. 3.2.5. The procedure implies that we solve the equations (2.5)

$$\begin{cases} F_n(x;c) = 0, \\ |F'_n(x;c) + 1| = 1 \end{cases} \tag{4.28}$$

by expansion near the point $(x, c) = (0, c_0)$, where c_0 is a root of the equation $F_k(0, c_0) = 0$. This means that we now write $x = 0 + \chi$, $c = c_0 + \varepsilon$ and assume that χ and ε are small – and this will be justified *a posteriori*. In this approximation

$$F_n(\chi, c) = c_n(c) - \chi + b_n \chi^d + O(\chi^{2d}),$$

and

$$c_{n+1} = c_n^d + c,$$
$$b_{n+1} = b_n c_n^{d-1} d.$$

At $c = c_0$ the $c_k(c_0) = 0$ and $c_k(c_0 + \varepsilon) = \dot{c}_k \varepsilon$ (dot denotes derivative with respect to c). Substituting $F_k = \dot{c}_k(c_0)\varepsilon - \chi + b_k(c_0)\chi^d + O(\chi^{2d})$ into (4.28), we get in this approximation:

$$db_k|\chi^{d-1}| = 1,$$
$$b_k\chi^d - \chi + \dot{c}_k\varepsilon = 0,$$

i.e. $\chi = \rho_k e^{i\phi}$, $\rho_k = (b_k d)^{-1/(d-1)}$ and

$$\partial\sigma_k : \quad \varepsilon = \frac{x - b_k x^d}{\dot{c}_k} = \frac{\rho_k}{\dot{c}_k}\left(e^{i\phi} - \frac{1}{d}e^{i\phi d}\right). \tag{4.29}$$

Thus such $\partial\sigma_k$ are approximately similar to $\partial S_1 = \partial\sigma_1$, described by Eq. (4.27).

Approximation works because the radius ρ_k is numerically small. Presumably it can be used to find Feigenbaum indices and their d-dependence. The same approximation can be used in the study of Julia sets for $c \in M_{k\alpha}$ with $k > 1$. From this calculation it is also clear that the shape of $M_{k\alpha}$ is dictated by the symmetry of the problem: for families μ of maps, which do not have any such symmetries one should expect all the higher $M_{k\alpha}$ to be described by (4.29) with $d = 2$. Obviously, Feigenbaum parameters depend only on effective value of d. Also, one can easily find families μ, for which the above approximated scheme will not work: then one can expect the breakdown of self-similarity of the Mandelbrot set.

However, if one tries to be more accurate, things get more sophisticated. Precise system (4.19) contains the condition $G_n(x;c) = 0$ rather than $F_n(x;c) = 0$ in (4.28). In variance with $F_n(x)$, which has a gap between x and x^d, $G_n(x)$ contains all powers of x, for example, for $n = 2$

$$G_2 = 1 + \sum_{k=0}^{d-1} c^{d-k-1}\chi^k + \sum_{k=0}^{d-2}(d-k-1)c^{d-k-2}\chi^{d+k} + \ldots + \chi^{d^{k-1}}.$$

This makes analysis more subtle. We leave this issue, together with approximate description of Julia sets to the future work.

4.10 Analytic case

Before ending the discussion of approaches to creation of theory of the universal Grand Mandelbrot (discriminant) variety and Julia sets and proceeding to examination of various examples, in the remaining two subsections we explain how the relevant notions look in the case of generic analytic functions, not obligatory polynomial. As already mentioned, the theory of discriminant, resultant and Mandelbrot varieties can be made pure topo-logical and free of any algebraic structure, however in this case one can lose other interesting – pure algebraic – examples (like $\mathbf{X} = \mathbf{F_p}$, to begin with) as well as the skeleton machinery, based on the use of functions F_n and G_k (to define them one needs subtraction operation and the entire ring structure respectively) and ordinary discriminant calculus (which is again algebraic). Thus in the remaining subsections we accept that these structures are important and explain instead, why we do *not* think that restrictions to polynomials are needed anywhere. Actually, polynomials play distinguished role in particular examples: this is because the vocabulary for iterations of other functions – even trigonometric – was never developed and we do not have any adequate language to discuss, say

$$G_2(x;\sin\omega x) = \frac{\sin(\omega\sin\omega x) - x}{\sin\omega x - x}$$

or

$$G_{1,1}(x;\sin\omega x) = \frac{\sin(\omega\sin\omega x) - \sin\omega x}{\sin\omega x - x} = G_2(x;\sin\omega x) - 1$$

or there no-less-interesting *trigonometric* counterparts

$$\tilde{G}_2(x;\sin\omega x) = \frac{\sin(\omega\sin\omega x) - \sin x}{\sin\omega x - \sin x},$$

$$\tilde{G}_{1,1}(x; \sin \omega x) = \frac{\sin(\omega \sin \omega x) - \sin \omega x}{\sin \omega x - \sin x} = \tilde{G}_2(x; \sin \omega x) - 1.$$

While iterated polynomial is always a polynomial, just of another degree, even iterated exponential or iterated sine do not have names. This could not be a too important restriction if one agrees to rely more on computer experiments than on theoretical considerations, but even here existing symbolic-calculus programs are better adjusted to work with polynomials. Still, non-polynomial singularities play increasingly important role in modern theories, they appear already in the simplest examples of τ-functions of KdV and KP equations, nothing to say about more general partition functions. Thus we find it important to emphasize that discussion of phase transitions and their hidden algebraic structure in the present book is by no means restricted to polynomials.

Take an arbitrary (locally) analytic function $f(x)$. By the same argument as in the polynomial case f maps roots of $F_n(x)$ into roots, giving us a representation of $\mathbf{Z_n}$ on the set of zeros of $F_n(x)$,[4] which is the analogue of the Galois group action for polynomials. The orders of orbits of this actions are also divisors of n, but the set of those orbits is now infinite for each given n. Although for series there is no notion of divisibility, still if n is divisible by k, then $F_k(x) = 0$ implies $F_n(x) = 0$. Indeed, if $n = km$ and $f^{\circ k}(x_0) = x_0$, then $f^{\circ n}(x_0) = f^{\circ(km)}(x_0) = f^{\circ(k(m-1))}(f^{\circ k}(x_0)) = f^{\circ(k(m-1))}(x_0) = \ldots = x_0$, i.e. every root of $F_k(x)$ is obligatory the root of $F_n(x)$.

Example: Take $f(x) = \sin(x) + cx + bx^4$. Figure 4.5 shows real roots of $F_4(x)$ and $F_2(x)$ for $c = 2$, $b = -0.01$. The root $x \approx 2.39$ is the root of $F_2(x)$ and its orbit has order 2. The root $x \approx 2.03$ is the root of G_4 (independent of $F_2(x)$) and its orbit has order 4.

If U is the definition domain (the set of regular points) of f, then $f \circ f$ and F_2 are defined in $U_{\circ 2} = U \cap f(U)$ and so on. Denote by

$$U^\infty(f) = \cap_{n=0}^\infty f^{\circ n}(U)$$

the inverse limit of the sequence

$$U \hookleftarrow U_{\circ 2} \hookleftarrow \ldots \hookleftarrow U_{\circ n} \hookleftarrow \ldots$$

For polynomial $f(x)$ the set $U^\infty(f) = \mathbf{C}$ is the whole complex plane.

---

[4]This is because if x_0 is a root of $F_n(x) = f^{\circ n}(x) - x$, then $f(x_0)$, $f^{\circ 2}(x_0)$ etc. are also roots, and $f^{\circ n}(x_0) = x_0$.

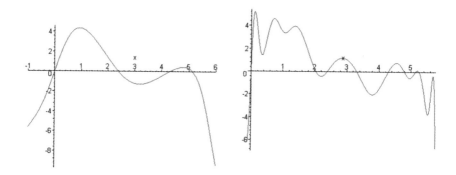

Fig. 4.5   Real roots of $F_2(x; f)$ and $F_4(x; f)$ for the map $f(x) = \sin(x) + 2x - 0,01x^4$.

The $\mathbf{Z_n}$ representation on the roots of F_n is also a representation of $\mathbf{Z}$, which we denote by V_n. If n is divisible by k, then every root of F_k is the root of F_n and we have an embedding, which is actually a morphism of $\mathbf{Z}$-representations $V_k \to V_n$. This provides a directed set of $\mathbf{Z}$-representations

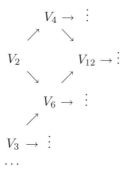

Its direct limit $V_\infty(f) = \cup_n V_n$ is a representation of $\mathbf{Z}$ which is a pertinent characteristic of the map f. All roots of every F_n satisfy $x = f^{\circ n}(x) = f^{\circ(nm)}(x)$ for any m, thus they belong to U^∞, i.e. each orbit of V_∞ totally belongs to U^∞. Thus we have the following hierarchy of subsets, characterizing a holomorphic function f:

$$V_\infty(f) \subset U^\infty(f) \subset \ldots \subset U_{\circ 2}(f) \subset U(f) \subset \mathbf{C}.$$

The closure of V_∞-space is called the **algebraic Julia set** J_A of f.

Hypothesis: This definition coincides with the one in Sec. 4.6.

Example: For $f(x) = x^d$ the Julia set $J(f)$ is the unit circle $S^1 \subset \mathbf{C}$ and the V_∞ space is an everywhere dense subset in this circle.

In the general case, when f is a series with complex coefficients, there is no distinguished f-invariant subsets analogous to $\mathbf{R} \subset \mathbf{C}$. But the existence of f-orbits implies that a deformation of coefficients of $f(x)$, which makes any two roots of some $F_n(x; f)$ lying on different orbits in $V_\infty(f) \subset U^\infty(f)$ coincide, results in simultaneous pairwise merging of *all* the roots on those orbits, so the resulting orbit acquires multiplicity.

For a given n take the subset C_n in the space of coefficients a of f where some of the orbits in the set of zeroes of F_n merge, i.e. f belongs to the I_n-pullback of the discriminant component $\mathcal{D}_{\mathrm{on}}$. The closure of $C_\infty := \cup_n C_n$ is the boundary ∂M of the **Mandelbrot set** of f.

Hypothesis: For any finite-parameter family $\mu \subset \mathcal{M}$ of maps $\overline{C_\infty}(\mu)$ is a subset of *real* codimension 1 in the family μ.

Provided this is true, ∂M separates the space of complex coefficients into disjoint components. Note that every particular C_n has real codimension 2 (since the two equations $F_n(x) = 0$ and $F_n'(x) = 0$ can be used to define a complex common root x and one complex relation between the coefficients of $F_n(x)$ – and thus between the coefficients of f).

4.11 Discriminant variety $\mathcal{D}$

This section gives the standard definitions of discriminants and resultants for the case of polynomial functions and briefly comments on the possible ways of generalization to the case of arbitrary analytic functions.

4.11.1 *Discriminants of polynomials*

In the polynomial case $\mathcal{D}(\mathcal{P}_n)$ is defined [13] as an algebraic variety in $\mathcal{P}_n$ by an equation

$$D(F) = 0 \qquad (4.30)$$

in the space of coefficients. Here $D(F)$ is the square of the Van-der-Monde product, for $F(x) = v_n \prod_{k=1}^n (x - \alpha_k)$

$$D(F) = v_n^{2n-2} \prod_{k<l} (\alpha_k - \alpha_l)^2. \qquad (4.31)$$

Remarkably, it can also be expressed as a **resultant** of $F(x)$ and its derivative $F'(x)$

$$D(F) = (-)^{n(n-1)/2} v_n^{-1} R(F(x), F'(x)), \qquad (4.32)$$

which is a polynomial of the coefficients of $F(x)$ (while the individual roots α_k can be sophisticated functions of these coefficients, not even expressible in radicals), so that equation (4.30) is indeed algebraic.

The resultant of two polynomials $F(x) = \sum_{k=0}^{n} v_k x^k = v_n \prod_{k=1}^{n}(x - \alpha_k)$ and $G(x) = \sum_{k=0}^{m} w_k x^k = w_m \prod_{l=1}^{m}(x - \beta_l)$ is defined as a double product

$$R(F,G) = v_n^m w_m^n \prod_{k,l}(\alpha_k - \beta_l) = v_n^m \prod_{k=1}^{n} G(\alpha_k) = (-)^{nm} w_m^n \prod_{l=1}^{m} F(\beta_l),$$

(4.33)

depends on symmetric functions of the roots $\{\alpha\}$ and/or $\{\beta\}$, thus – according to Vieta formula – depends only on the coefficients of F and/or G, and is representable as determinant

$$R(F,G) = \det_{(m+n) \times (m+n)} \begin{pmatrix} v_0 & v_1 & \dots & v_n & & & \\ & v_0 & v_1 & \dots & v_n & & \\ & & \dots & & & & \\ & & v_0 & v_1 & \dots & v_n & \\ w_0 & w_1 & \dots & w_m & & & \\ & w_0 & w_1 & \dots & w_m & & \\ & & \dots & & & & \\ & & w_0 & w_1 & \dots & w_m & \end{pmatrix}.$$

(4.34)

The resultant obviously vanishes when $F(x)$ and $G(x)$ have a common root x_0, accordingly the matrix in (4.34) annihilates the column-vector $(1, x_0, x_0^2, \dots, x_0^{n+m})$.

Directly from the definitions it follows that discriminant of a product of two polynomials is decomposed in the following way

$$D(FG) = D(F)D(G)R^2(F,G).$$

(4.35)

If $F(x) - H(x)$ is divisible by $G(x)$, $F = H \mod G$ (for example, H is the residual from F division by G), then (4.33) implies that

$$R(F,G) \sim \prod_{\beta:\, G(\beta)=0} F(\beta) \sim \prod_{\beta:\, G(\beta)=0} H(\beta) \sim R(H,G) \sim \prod_{\gamma:\, H(\gamma)=0} G(\gamma)$$

(4.36)

where proportionality signs imply the neglect of factors like v_n, w_m and minus signs in (4.33), in this form the relation is applicable to any analytic functions, not obligatory polynomials. (Note that the products on the r.h.s. of (4.36) over the roots of H and, say, of $H + \kappa G$ will be the same – up to the above-mentioned rarely essential factors.)

4.11.2 *Discriminant variety in entire* $\mathcal{M}$

To define the discriminant in analytic (non-polynomial) case one can make use of both definitions (4.31) and (4.32) and either take the limit of a double-infinite product over the roots or handle determinant of an infinite matrix. It is an open question, when exactly this can be done and when the two limits can coincide.

A possible approach in the case of determinantal representation can make use of the following recursive procedure. Discriminant of a polynomial $a_0 + a_1 x + \ldots + a_{d+1} x^{d+1}$ may be written as (pre-factor a_{d+1}^{-1} in (4.32) is important here)

$$D_{d+1}\{a_0, \ldots, a_{d+1}\} = a_d^2 \cdot D_d\{a_0, \ldots, a_d\} + \sum_{k=1}^{d} b_{d+1,k}\{a_0, \ldots, a_d\} \cdot a_{d+1}^k,$$

where $D_d\{a_0, \ldots, a_d\}$ is discriminant of $a_0 + a_1 x + \ldots + a_d x^d$ of the previous degree and $b_{d+1,k}\{a_0, \ldots, a_d\}$ are polynomials of coefficients $a_0, \ldots, a_d$. Then for an analytic function locally represented by series $f(x) = \sum a_i x^i$ with numerically given coefficients $\{a_i\}$ we have a sequence

$$\check{D}_1 = 1, \quad \check{D}_2 = a_1^2 - 4a_0 a_2,$$
$$\check{D}_3 = a_2^2(a_1^2 - 4a_0 a_2) + (18a_0 a_1 a_2 - 4a_1^3)a_3 - 27a_0^2 a_3^2,$$
$$\ldots, \check{D}_d, \ldots$$

of numbers, equal to the values of the corresponding polynomials. The numeric limit of this sequence may be regarded as the value of the infinite expression

$$\ldots (a_d^2 \cdot \ldots \cdot (a_3^2 \cdot (a_2^2 \cdot (a_1^2 \cdot 1 + b_{21}\{a_0, a_1\}a_2)$$

$$+ b_{31}\{a_0, a_1, a_2\}a_3 + b_{32}\{a_0, a_1, a_2\}a_3^2)$$

$$+ b_{41}\{a_0, a_1, a_2, a_3\}a_4 + b_{42}\{a_0, a_1, a_2, a_3\}a_4^2 + b_{43}\{a_0, a_1, a_2, a_3\}a_4^3)$$

$$+ \cdots + \sum_{i=1}^{d} b_{d+1,i}\{a_0, \ldots, a_d\}a_{d+1}^i) \ldots$$

(this computation is an algebraic counterpart of the chain fractions) which may be called the **discriminant** $D(f)$ of the analytic function f.

Since in the case of polynomial f the coefficients of all functions F_n are polynomials $\{a_k^{(n)}(a)\}$ of the coefficients $\{a_k\}$ of the series f, such procedure can work simultaneously for f and $F_n(f)$. The possibility of generalizations from polynomial to analytic case can be formulated in the form of the following

Hypothesis: $\{a_i\} \in C_n$ iff $D(F_n) = 0$, i.e. some orbits within the radius of convergence of series $f = \sum a_i x^i$ merge iff the above defined discriminant $D(F_n)$ of series $F_n = \sum a_k^{(n)}(a)x^k$ is zero.

Example: For $f_c(x) = e^{x^2} - c$ we have $\check{D}_1 = 1$, $\check{D}_2 = 4(c - 1)$, $\check{D}_3 = 4(c - 1)$, $\check{D}_4 = 8(c - 1)(2c - 1)^2$, ... It is easy to check that all higher $\check{D}_k$ will also be proportional to $c - 1$, so that discriminant $D\left(e^{x^2} - c\right)$, defined as numerical limit of this sequence vanishes at $c = 1$, where the two real roots of $f_c(x)$ merge (and go to complex domain).

4.12 Discussion

The above considerations suggest the following point of view.

Given a (complex) analytic function $f(x)$ we get in the domain $U^\infty(f) \subset \mathbf{X}$ a set of discrete f-invariant indecomposable subsets. These subsets are orbits of a **Z**-action on $U^\infty(f)$, generated by f. The union of periodic orbits gives a **Z**-representation $V_\infty(f)$ which may be regarded as an intrinsic characteristic of f. Each periodic orbit belongs to a set of roots of some $G_n(x; f)$ of minimal degree n, which can be assigned to the orbit as its degree. As the shape of f changes, i.e. as f moves in the moduli space $\mathcal{M}$, the f-orbits move in $\mathbf{X}$. The dynamics of this motion is worth studying. For f's of certain shapes some orbits can merge. The corresponding subset in $\mathcal{M}$ can be regarded as inverse image (pullback) $\mathcal{D}^*$ of the discriminant variety $\mathcal{D} \subset \mathcal{M}$, induced by the functions $G_n(x; f)$. We believe that discriminants and resultants can be well-defined not only for polynomials, but also for analytic functions, while the variety $\mathcal{D}^*$ makes sense in even more general – topological – setting. We identify the closure of the union $\cup_{m,n=1}^\infty \mathcal{R}_{mn}^* = \cup_{n=1}^\infty \mathcal{D}_n^*$ as the boundary of the Mandelbrot set, which separates the moduli space $\mathcal{M}$ into disjoint components, which can be used to classify analytic functions. Even more structures (Julia sets, secondary and Grand Mandelbrot sets) arise in a similar way from consideration of pre-orbits and grand orbits of f and of their change with the variation of f.

Even in multi-dimensional case for every map $\vec{f}(\vec{x}) \in \mathcal{M}$ one can take its iterations $\vec{F}_n(\vec{x}; \vec{f}) = f^{\circ n}(\vec{x}) - \vec{x}$ and construct the real-codimension-two subsets $\mathcal{D}_n^*$ and $\mathcal{R}_{mn}^*$ in $\mathcal{M}$, made out of zeroes of reduced discriminant $d_n(\vec{f})$ and resultant $r_{mn}(\vec{f})$ functions.[5] A union $\bigcup_{m=1}^{\infty} \mathcal{R}_{nm}^* = \partial S_n$ has real codimension one in $\mathcal{M}$ and can be considered as a boundary of a codimension-zero *stability set* $S_n \subset \mathcal{M}$, so that $d_n \in \partial S_n$ and $\partial S_n \cap \partial S_m = \mathcal{R}_{mn}^*$. What is non-trivial, the intersection of stability domains, not only their boundaries, is just the same (and has real codimension two!): $\partial S_n \cup \partial S_m = \mathcal{R}_{mn}^*$. This makes the structure of *universal Mandelbrot set* $M = \cup_n S_n$ non-trivial, and requires description in terms of projections to *multipliers trees* and representation theory, as suggested in this book. Sections of the Mandelbrot set by real-dimension-two manifolds (by one-complex-parametric map families $\mu_c \subset \mathcal{M}$) help to visualize some of the structure, but special care should be taken to separate intrinsic properties of M from peculiarities of particular section $M(\mu) = M \cup \mu$. For example, the tree structure is universal, while the number of vertices and links is not.

---

[5]The function d_n can be defined as irreducible constituent of

$$D_n = \begin{cases} \vec{F}_n(\vec{x}; \vec{f}) = 0 \\ \det_{ij} \frac{\partial F_n^i}{\partial x_j}(\vec{x}; \vec{f}) = 0, \end{cases}$$

and $r_{mn} = 0$ whenever $\vec{F}_m(\vec{x}; \vec{f}) = \vec{F}_n(\vec{x}; \vec{f}) = 0$ (though seemingly overdefined, this system defines a real-codimension-two subspace in $\mathcal{M}$ if $m|n$ or $n|m$). Stability set S_n can be defined as

$$S_n = \begin{cases} \vec{F}_n(\vec{x}; \vec{f}) = 0 \\ \left| \text{e.v.} \left(\delta_j^i + \frac{\partial F_n^i}{\partial x_j}(\vec{x}; \vec{f}) \right) \right| < 1, \end{cases}$$

where e.v.(A_j^i) denote eigenvalues of matrix A.

Chapter 5

Map $f(x) = x^2 + c$: from standard example to general conclusions

In this section we use the well-publicized example to illustrate our basic claims. Namely:

1) The pattern of periodic-orbit bifurcations in *real* case is well-described in the language of discriminants.

2) Discriminants carry more information in two aspects: there are more bifurcations in real case than captured by the period-doubling analysis and there are even more bifurcations in complex domain which are not seen in real projection.

3) Fractal Mandelbrot set is in fact a union of well-defined domains with boundaries, which (i) are described by algebraic equations, (ii) are smooth almost everywhere (outside zeroes of the associated discriminants), (iii) are densely populated by zeroes of appropriate resultants. When resultant vanishes, a pair of smooth domains touch and the touching point serves as a single "bridge", connecting the two domains.

4) Julia set is related to inverse limit of almost any grand orbit: its infinitely many branches originate in the vicinities of infinitely many unstable orbits, which constitute the boundary of the algebraic Julia set.

5) Changes of stability (bifurcations of Julia set) always occur when something happens to periodic orbits, but these orbits that matter are not necessarily stable. Unstable orbits can be best studied with the help of grand orbits, because even pre-images of stable orbits carry information about unstable ones: all grand orbits originate in the vicinity of unstable orbits (grand orbit's periodic branches are more-or-less in one-to-one correspondence with unstable orbits). At the same time, everything what one can wish to know about grand orbits is again describable in terms of discriminants and resultants.

6) If $f \in M_1$, the fractal Julia set $J(f)$ is in fact a continuous (but

not necessarily smooth) deformation of a unit disc. The unit circle – the boundary of the disc – is densely covered by grand orbits, associated with unstable periodic orbits, and one stable periodic orbit, together with its grand orbit, lies inside the disc. The vanishing of the relevant resultant implies that the stable grand orbit from inside comes to the boundary, intersects and exchange stability with a particular unstable grand orbit, which becomes stable, quits the boundary and immerses into the Julia set. This results in merging the corresponding points of the boundary and changes the topology of Julia set – a bifurcation occurs. Topology of the Julia set $J(f)$ is defined by position of the map f in the Mandelbrot set and by the path through "bridges", which connects f with the central domain of the Mandelbrot set (i.e. by the branch of the powerful tree T_1).

7) If $f \in M_{k\alpha}$ with $k > 1$ the disc/ball is substituted by k-dependent collection of disjoint discs/balls, grand orbits are jumping between the discs, but return to the original one every n times, if the orbit is of order n. All bifurcations when f moves between different components inside the given $M_{k\alpha}$ are described in the same way as in 6). If f leaves Mandelbrot set, $f \notin M$, there are no stable orbits left and Julia set looses its "body" (interior of the disc), only the boundary survives, which contains all the bounded grand orbits.

5.1   Map $f(x) = x^2 + c$. Roots and orbits, real and complex

$f_c(x) = x^2 + c$ is a well-known example, examined in numerous papers and textbooks. This makes it convenient for illustration of general arguments.

5.1.1   *Orbits of order one (fixed points)*

• $\mathcal{S}_1(f_c)$ – is the set of roots of

$$F_1(x; f_c) = x^2 - x + c = G_1(x; f_c).$$

Since the order of polynomial f_c is $d = 2$, the number of roots of $F_1 = G_1$ is two, each of the two roots,

$$\mathcal{S}_1(f_c) = \left\{ \frac{1}{2} \pm \frac{\sqrt{1 - 4c}}{2} \right\}, \tag{5.1}$$

is individual orbit of order one.

The orbits = roots are real, when $c \leq \frac{1}{4}$.

The action of $\hat{f}(1)$ leaves each of the orbits = roots intact. The action of $\hat{f}(1)$ can be lifted to that of the entire Abelian group $\mathbf{Z}$, the same for all periodic orbits of arbitrary order (see Sec. 4.10). This action, however, does not distinguish between conjugate and self-conjugate orbits, to keep this information an extra $\mathbf{Z_2}$ group is needed. If we denote by v_q representation of $\mathbf{Z}$ when the group acts as cyclic permutations on the sequence of q elements, and label the trivial and doublet representations of $\mathbf{Z_2}$ by superscripts 0 and $\pm$, then the representation of $\mathbf{Z} \otimes \mathbf{Z_2}$ associated with $F_1(x)$ is $2v_1^0 = v_1^+ \oplus v_1^-$.

Discriminant $d(G_1) = D(G_1) = D(F_1(f_c)) = 1 - 4c$, its only zero is at $c = \frac{1}{4}$, where the two order-one orbits intersect. At intersection point the orbits are real, $x = \frac{1}{2}$ and remain real for real $c < \frac{1}{4}$. On the real line it looks like the two real roots are "born from nothing" at $c = \frac{1}{4}$.

Since $f'(x) = 2x$ the plus-orbit is unstable for all real c, the minus-orbit is stable on the segment $-3/4 < c < 1/4$.

In the domain of complex c stability region S_1 is bounded by the curve $|2x| = 1$ where x is taken from (5.1), i.e.

$$\partial S_1 : \quad \begin{array}{l} |1 \pm \sqrt{1-4c}| = 1, \quad \text{or} \\ c = 1/4 + e^{i\varphi}\sin^2\varphi/2. \end{array} \tag{5.2}$$

In polar coordinates (r, φ) with the center at $\frac{1}{4}$, the curve is given by $r = \sin^2\frac{\varphi}{2}$, see Fig. 5.1. Our expectation is that the points on this curve (an everywhere dense countable subset in it) will be zeroes of the resultants $R(F_1, F_n)$, or $r(G_1, G_n)$ to be precise, with all possible n.

The critical point of the map $f_c = x^2 + c$, i.e. solution to $f_c'(x) = 0$ is $w_c = 0$. Equation $F_1(w_c; f_c) = F_1(0; f_c) = f_c(0) = 0$ has a single solution, $c = 0$, which lies inside the single elementary component of $S_1 = \sigma^{(0)}[1]$.

5.1.2 *Orbits of order two*

• $S_2(f_c)$ – the roots of

$$F_2(x; f_c) = (x^2 - x + c)(x^2 + x + c + 1) = G_1(x; f_c)G_2(x; f_c). \tag{5.3}$$

F_2 is divisible by F_1 since 1 is a divisor of 2, the ratio

$$G_2(x; f_c) = F_2/F_1 = x^2 + x + c + 1.$$

The $d^2 = 4$ roots of F_2 form two conjugate order-one orbits and one ($\frac{d^2-d}{2} = 1$) self-conjugate order-two orbit. The new orbit – in addition to the two order-one orbits inherited from $S_1(f_c)$ – is of order 2 and consists of two points $-\frac{1}{2} \pm \frac{\sqrt{-3-4c}}{2}$ – the zeroes of G_2.

The action of $\hat{f}(2)$ interchanges these two points. Representation of $\mathbf{Z} \otimes \mathbf{Z_2}$, associated with F_2, is $2v_1 \oplus v_2 = (v_1^+ \oplus v_1^-) \oplus v_2^0$.

Discriminant

$$D(F_2) = (4c-1)(4c+3)^3$$
$$= D(G_1)D(G_2)R^2(G_2, G_1) = d(G_1)d^2(G_2)r^3(G_2, G_1),$$

and

$$D(G_1) = d(G_1) = 1 - 4c,$$
$$D(G_2) = d^2(G_2)R(G_2, G_1) = -3 - 4c,$$
$$d(G_2) = -1,$$
$$R(G_2, G_1) = r(G_2, G_1) = 3 + 4c.$$

The only new zero is at $c = -\frac{3}{4}$. Since it is a zero of $R(G_2, G_1)$, the new orbit, associated with the roots of $G_2(x; f_c)$, intersects an old one, related to zeroes $G_1(x; f_c)$. Indeed, at $c = -3/4$ the stable order-one orbit intersects the order-2 orbit and loses stability, while the order-2 orbit becomes stable and stays real for all real $c < -\frac{3}{4}$.

Since at intersection point $c = -\frac{3}{4}$ the two roots of G_2 merge, $D(G_2)$ has a simple zero. Similarly, below all zeroes of $R(G, \tilde{G})$ will also be zeroes of either $D(G)$ or $D(\tilde{G})$ (in particular, $D(G)$ are usually reducible polynomials over any field $\mathbf{X}$), moreover their multiplicities are dictated by the properties of intersecting orbits.

Stability criterium for the order-2 orbit of $f_c(x)$ is

$$\begin{cases} |f'(x)f'(f(x))| = |4xf_c(x)| = |4x(x^2 + c)| < 1 \\ G_2(x; f_c) = x^2 + x + c + 1 = 0. \end{cases}$$

The boundary of stability region S_2 in complex plane c is obtained by changing inequality for equality. Then $x(x^2 + c) = -x(x + 1) = c + 1$, and the boundary is just a circle

$$\partial S_2 : \quad |c + 1| = \frac{1}{4} \tag{5.4}$$

with center at $c = -1$ and radius $\frac{1}{4}$, see Fig. 5.1. It intersects with the curve (5.2) at $c = -\frac{3}{4}$, which is the zero of $R(G_2, G_1) = 3 + 4c$.

$G_2(w_c; f_c) = G_2(0; c) = c + 1 = 0$ now implies that $c = -1$, and this point lies inside the single elementary domain of $S_2 = \sigma^{(1)}[2|1]$.

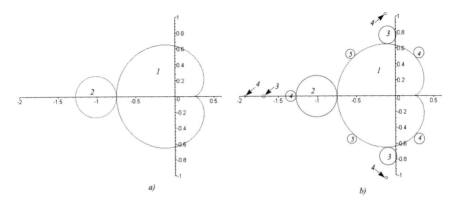

a)

b)

Fig. 5.1   Stability domains in the plane of complex c: a) for orbits of order one and two, b) for orbits of orders up to four.

5.1.3   Orbits of order three

• $\mathcal{S}_3(f_c)$ – the roots of

$$F_3(x; f_c) = (x^2 - x + c)\left(x^6 + x^5 + x^4(3c + 1) + x^3(2c + 1)\right.$$
$$+ x^2(3c^2 + 3c + 1) + x(c^2 + 2c + 1) + (c^3 + 2c^2 + c + 1)\big)$$
$$= G_1(x; f_c)G_3(x; f_c).$$

$F_3(x)$ is divisible by $F_1(x) = G_1(x)$ since 1 is a divisor of 3, the ratio $G_3(x) = F_3(x)/F_1(x)$.

The $d^3 = 8$ roots of F_3 form two conjugate order-one orbits and two $(\frac{d^3 - d}{3} = 2)$ conjugate order-3 orbits.

Representation of $\mathbf{Z} \otimes \mathbf{Z_2}$, associated with F_3 is $2v_1 \oplus 2v_3 = (v_1^+ \oplus v_1^-) \oplus (v_3^+ \oplus v_3^-)$.

Discriminant

$$D(F_3) = (4c - 1)(4c + 7)^3(16c^2 + 4c + 7)^4$$
$$= D(G_1)D(G_3)R^2(G_3, G_1) = d(G_1)D^3(G_3)r^4(G_3, G_1)$$

and

$$D(G_1) = d(G_1) = 1 - 4c,$$
$$D(G_3) = d^3(G_1)r^2(G_3, G_1) = -(4c + 7)^3(16c^2 + 4c + 7)^2,$$
$$d(G_3) = -(4c + 7),$$
$$R(G_3, G_1) = r(G_3, G_1) = 16c^2 + 4c + 7.$$

The new – as compared to F_1 (but not F_2, which is not a divisor of F_3) – zeroes of $D(F_3)$ are at $c = -\frac{7}{4}$ and at $c = -\frac{1}{8} \pm \frac{3\sqrt{3}i}{8}$. The latter two values are zeroes of $R(G_3, G_1)$ and they describe intersection of two order-3 orbits with the stable order-1 orbit (one of the two roots of G_1) at essentially complex c. No trace of these intersections is seen in the plane of real x and c. The two order-3 orbits intersect at the remaining zero of $D(F_3)$, which is real, $c = -\frac{7}{4}$, and they (orbits) remain real at all real $c < -\frac{7}{4}$. However, since $c = -\frac{7}{4}$ is actually a root of $d(G_3)$, it has nothing to do with G_1, and the new real order-3 orbits look – from the *real* point of view – "born from nothing", not at the position of any other "previously existing" orbits. In complex domain the orbits always exist, and just intersect (coincide) get real at $c = -\frac{7}{4}$.

Stability criterium for an order-3 orbit of $f_c(x)$ is

$$\begin{cases} |f'(x)f'(f(x))f'(f^{\circ 2}(x))| = |8xf_c(x)f_c^{\circ 2}(x)| \\ \quad = |8x(x^2 + c)(x^4 + 2cx^2 + c^2 + c)| < 1, \\ G_3(x; c) = 0. \end{cases}$$

This system is hard to solve explicitly. Solving instead

$$G_3(0, c) = c^3 + 2c^2 + c + 1 = 0$$

we obtain three points, lying inside the three elementary domains of

$$S_3 = \sigma^{(0)}[3] \bigcup \sigma^{(1)} \begin{bmatrix} 3 \\ 1 \end{bmatrix} 1 \bigcup \sigma^{(1)} \begin{bmatrix} 3 \\ 2 \end{bmatrix} 1 .$$

This is the first example when we observe the appearance of α-parameters.

5.1.4   *Orbits of order four*

- $S_4(f_c)$ – the roots of

$$F_4(x; f_c) = (x^2 - x + c)(x^2 + x + c + 1)\left(x^{12} + 6cx^{10} + x^9\right.$$
$$+ x^8(15c^2 + 3c) + 4cx^7 + x^6(20c^3 + 12c^2 + 1) + x^5(6c^2 + 2c)$$
$$+ x^4(15c^4 + 18c^3 + 3c^2 + 4c) + x^3(4c^3 + 4c^2 + 1)$$
$$+ x^2(6c^5 + 12c^4 + 6c^3 + 5c^2 + c) + x(c^4 + 2c^3 + c^2 + 2c)$$
$$\left. + (c^6 + 3c^5 + 3c^4 + 3c^3 + 2c^2 + 1)\right)$$
$$= F_2(x)G_4(x) = G_1(x)G_2(x)G_4(x).$$

F_4 is divisible by F_1 and F_2 since 1 and 2 are divisors of 4, the ratio $G_4 = F_4/F_2$ is a polynomial of order 12.

The $d^4 = 16$ roots of F_4 form two conjugate order-one orbits (inherited from G_1), one self-conjugate order-2 orbit (from G_2) and three ($\frac{d^4 - d^2}{4} = 3$) new order-4 orbits, two conjugate and one self-conjugate.

Representation of $\mathbf{Z} \otimes \mathbf{Z}_2$, associated with F_4, is

$$2v_1 \oplus v_2 \oplus 3v_4 = (v_1^+ \oplus v_1^-) \oplus v_2^0 \oplus v_4^0 \oplus (v_4^+ \oplus v_4^-).$$

Discriminant

$$D(F_4) = (4c - 1)(4c + 3)^3$$
$$\cdot (4c + 5)^6 (16c^2 - 8c + 5)^5 (64c^3 + 144c^2 + 108c + 135)^4$$
$$= D(G_1)D(G_2)D(G_4)R^2(G_2, G_1)R^2(G_4, G_1)R^2(G_4, G_2)$$
$$= d(G_1)d^2(G_2)d^4(G_4)r^3(G_2, G_1)r^5(G_4, G_1)r^6(G_4, G_2)$$

where

$$D(G_1) = 1 - 4c,$$
$$D(G_2) = -3 - 4c,$$
$$D(G_4) = (4c + 5)^2(16c^2 - 8c + 5)^3(64c^3 + 144c^2 + 108c + 135)^4,$$
$$R(G_2, G_1) = 3 + 4c,$$
$$R(G_4, G_1) = (16c^2 - 8c + 5),$$
$$R(G_4, G_2) = (4c + 5)^2$$

and

$$d(G_1) = 1 - 4c,$$
$$d(G_2) = -1,$$
$$d(G_4) = 64c^3 + 144c^2 + 108c + 135,$$
$$r(G_2, G_1) = 3 + 4c,$$
$$r(G_4, G_1) = 16c^2 - 8c + 5,$$
$$r(G_4, G_2) = 4c + 5.$$

The roots of $D(F_4)$ describe the following phenomena:

$c = \frac{1}{4}$ and $c = -\frac{3}{4}$ were already examined in connection with F_1 and F_2.

$c = -\frac{5}{4}$ is a zero of $D(G_4)$, thus it could describe either intersection of two order-4 orbits or merging of roots of a single order-4 orbit. In the latter case that orbit would degenerate into the one of order 2 or 1. Since this c is simultaneously the zero of $R(G_4, G_2)$, the right choice is intersection of self-conjugate order-4 orbit with the previously existing (inherited from

G_2) self-conjugate order-2 orbit. Since at intersection point two pairs of roots of G_4 merge, the order of zero of $D(G_4)$ is two. For lower real $c < -\frac{5}{4}$ the self-conjugate order-4 orbit remains real.

$c = \frac{1}{4} \pm \frac{i}{2}$ are complex zeroes of $R(G_4, G_1)$ and describe intersection of conjugate order-1 orbits with the conjugate order-4 orbits in complex domain. No traces of this intersection are seen in the plane of real x and c. Since at intersection points all the four roots of G_4 merge together, $D(G_4)$ has cubic zeroes.

The remaining three roots of $D(F_4)$ are roots of $d(G_4)$ alone and describe entirely the world of order-4 orbits: the $C_3^2 = 3$ intersections of 3 orbits. Namely,

$c = -\frac{3}{4}(1 + 2^{2/3})$ – two conjugate order-4 orbits intersect, get real and stay real for lower real values of c. Look as "born from nothing" from the real perspective. Since 4 pairs of roots need to merge simultaneously, the order of zero of $D(G_4)$ is four.

$c = -\frac{3}{4} + \frac{3}{8}2^{2/3}(1 \pm i\sqrt{3})$ – self-conjugate order-4 orbit intersects with one or another of the two conjugate order-4 orbits in complex domain. No traces at real plane. $D(G_4)$ has quadruple zero.

Stability criterium for an order-4 orbit of $f_c(x)$ is

$$
\begin{cases}
|f'(x)f'(f(x))f'(f^{\circ 2}(x))f'(f^{\circ 3}(x))| = |2^4 x f_c(x) f_c^{\circ 2}(x) f_c^{\circ 3}(x)| \\
\quad = |8x(x^2 + c)(x^4 + 2cx^2 + c^2 + c)(x^8 + \ldots)| < 1, \\
G_4(x; c) = 0.
\end{cases}
$$

Solving instead of this system

$$
G_4(0, c) = c^6 + 3c^5 + 3c^4 + 3c^3 + 2c^2 + 1 = 0
$$

we obtain six points, lying inside the six elementary domains of

$$
S_4 = \left(\bigcup_{\alpha=1}^{3} \sigma^{(0)}[4, \alpha] \right) \bigcup \sigma^{(1)} \begin{bmatrix} 4 \\ 1 \end{bmatrix} 1 \end{bmatrix} \bigcup \sigma^{(1)} \begin{bmatrix} 4 \\ 2 \end{bmatrix} 1 \end{bmatrix} \bigcup \sigma^{(2)}[2\ 2|1].
$$

5.1.5 *Orbits of order five*

• $S_5(f_c)$ – the roots of

$$
F_5(x; f_c) = (x^2 - x + c)(x^{30} + \ldots) = G_1(x)G_5(x).
$$

There are $\frac{d^5 - d}{5} = 6$ order-5 orbits.

$$D(G_5) = d^5(G_5)r^4(G_1, G_5),$$
$$R(G_1, G_5) = r(G_1, G_5) = 256c^4 + 64c^3 + 16c^2 - 36c + 31,$$
$$d(G_5) = -(4194304c^{11} + 32505856c^{10}$$
$$+109051904c^9 + 223084544c^8 + 336658432c^7$$
$$+492464768c^6 + 379029504c^5 + 299949056c^4 + 211327744c^3$$
$$+120117312c^2 + 62799428c + 28629151)$$
$$= -\left((4c)^{11} + 31(4c)^{10} + 416(4c)^9 + \ldots\right). \quad (5.5)$$

The powers 4 and 5 appear here because when an order-5 orbit intersects with the order-1 orbit, the 5 different roots of F_5 should merge simultaneously, thus discriminant has a zero of order $4 = 5 - 1$, while at intersection of any two order-5 orbits, the 5 pairs of roots should coincide pairwise, so that discriminant has a zero of order 5.

$G_5(0; c)$ is polynomial of order $2 \cdot 8 - 1 = 15$ in c, thus equation $G_5(0; c) = 0$ has 15 solutions. These should describe the would be $C_6^2 = 15$ pairwise intersections of 6 order-5 orbits. However, since the power of c in $d(G_5)$ is only 11, we conclude that only 11 of such intersections really occur. Instead there are 4 intersections between order-1 and order-5 orbits (since $r(G_1, G_5)$ is of order 4 in c).

$$15 = 11 + 4,$$

and stability domain S_5 consists of 15 elementary domains.

5.1.6 Orbits of order six

- $S_6(f_c)$ – the roots of

$$F_6(x; f_c) = G_1(x)G_2(x)G_3(x)G_6(x) =$$
$$(x^2 - x + c)(x^2 + x + c + 1)(x^6 + \ldots)(x^{54} + \ldots).$$

Here the degree-54 polynomial $G_6 = F_6 F_1 / F_2 F_3 = F_6 / F_1 (F_2 / F_1)(F_3 / F_1)$ since 6 has three divisors 1, 2 and 3. There are $\frac{d^6 - d^2 - d^3 + d}{6} = 9$ order-6 orbits.

$$R(G_1, G_6) = r(G_1, G_6) = 16c^2 - 12c + 3,$$
$$R(G_2, G_6) = r^2(G_2, G_6) = (16c^2 + 36c + 21)^2,$$
$$R(G_3, G_6) = r^3(G_3, G_6) = (64c^3 + 128c^2 + 72c + 81)^3,$$
$$D(G_6) = -d^6(G_6)R^5(G_1, G_6)R^2(G_2, G_6)R(G_3, G_6)$$
$$= d^6(G_6)r^5(G_6, G_1)r^4(G_6, G_2)r^3(G_6, G_3),$$
$$d(G_6) = 1099511627776c^20 + 10445360463872c^19 + 44873818308608c^18$$
$$+121736553037824c^17 + 245929827368960c^16 + 399107688497152c^15$$
$$+535883874828288c^14 + 617743938224128c^13 + 631168647036928c^12$$
$$+576952972869632c^11 + 484537901514752c^10 + 376633058918400c^9$$
$$+263974525796352c^8 + 173544017002496c^7 + 104985522188288c^6$$
$$+58905085704192c^5 + 33837528259584c^4 + 15555915962496c^3$$
$$+8558772746832c^2 + 1167105374568c + 3063651608241 =$$
$$= (4c)^{20} + \ldots$$

$$(5.6)$$

$G_6(0, c)$ is polynomial of degree $32 - 1 - 1 - 3 = 27$ in c: there are 27 solutions to $G_6(0, c) = 0$. On the other hand there could be up to $C_9^2 = 36$ pairwise intersections between 9 order-6 orbits, however there are only 20, instead there are 2 intersections between order-6 and order-1, 2 – between order-6 and order-2, and 3 - between order-6 and order-3 orbits (all these numbers are read from the powers of c in the reduced discriminant $d(G_5)$ and reduced resultants $r(G_6, G_i)$). We have

$$27 = 20 + 2 + 2 + 3,$$

stability domain S_6 has 27 elementary components.

In this subsection we used the example of $f_c(x) = x^2 + c$ to demonstrate the application of discriminant analysis. We explained in what sense the standard period-doubling bifurcation tree in Fig. 5.2 is a part of a more general pattern. Even on the plane of real x and c much more is happening. Period doublings do not exhaust all possible bifurcations, other orbits of various orders are "born from nothing" – not at the positions of previously existing orbits – if we move from higher to lower c or "merge and disappear" if we move in the opposite direction. If we go away from real c and x, period doubling gets supplemented by tripling, quadrupling and so on. In complex domain no orbits can be born, merge or disappear – they can only intersect in different ways. Moreover, such intersections do not need to leave any traces on the real $c - x$ plane. Specifics of the iterated maps F_n is that whenever they hit the discriminant variety the whole orbits intersect: many different roots coincide simultaneously. As n increases, the pattern becomes more and more sophisticated and begins looking chaotic as $n \to \infty$. Still the whole story is exactly that of the intersecting algebraic varieties (infinite-dimensional, if we speak about infinitely-large n). We used the standard example in order to formulate the appropriate language for discussion of these – in fact, pure algebraic – phenomena.

5.2 Mandelbrot set for the family $f_c(x) = x^2 + c$

Figure 5.1 shows stability domains for different orbits (roots of the polynomials F_n and G_n) in the complex c plane. Figure 5.2 shows the full picture, where one can easily recognize the standard Mandelbrot set [9]. We see that it is nothing but the union of stability domains for orbits of different orders, which touch at single points: at zeroes of associated resultants. The picture also shows in what sense the boundary of every particular stability domain is formed by the zeroes of resultants (the zeroes constitute count-able dense subsets in these boundaries). Note also that the two domains touch only if the order of one orbit divides another, otherwise the resul-tants do not have zeroes. For example, $R(G_2, G_3) \equiv 1$, and it is easy to see that indeed these two r-polynomials cannot have a common zero, since $G_3(x; f_c) - (x^4 + 2cx^2 + x + c^2 + c)G_2(x; f_c) = 1$. Similarly, $R(G_4, G_6) \equiv 1$ etc.

According to this picture every point in the Mandelbrot set (i.e. every map f from a given family – not obligatory one-parametric) belongs to one particular stability domain, and there is a uniquely defined sequence of "bridges" (tree structure), which should be passed to reach the "central" stability domain – the one with a stable fixed point. As we shall see below, this sequence (path) is the data, which defines the structure of the Julia set $J(f)$. As will be demonstrated in Chap. 6, from the point of view of this picture there is nothing special in the maps $f_c(x) = x^2 + c$, neither in quadraticity of the map, nor in unit dimension of the family: the *tree-forest-trail* structure of Mandelbrot set is always the same.

However, particular numbers of vertices (elementary domains σ), links (zeroes of r_{mn}'s), cusps (zeroes of d_n's) and trails are *not* universal – depend

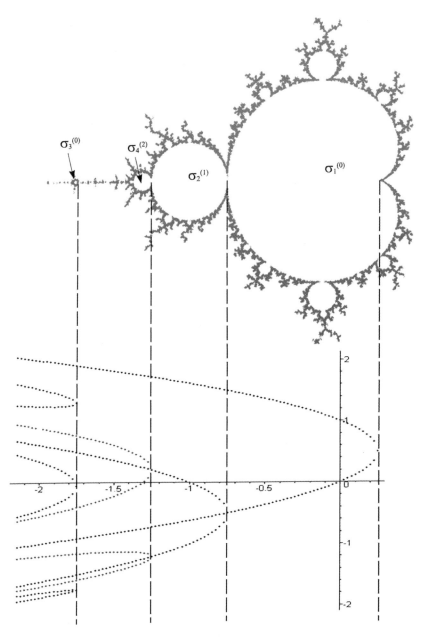

Fig. 5.2   Mandelbrot set = union of all stability domains in the plane of complex c for the family of maps $f_c(x) = x^2 + c$. Boundary of the white region is densely filled by collection of all zeroes of all resultants = discriminant variety. The picture below shows the real section of the "Julia sheaf", by the plane of real x and real c: for each c on the real line shown are the real orbits of different orders, stable and unstable.

on the choice of the family μ of maps (on the section of the universal Mandelbrot set). Moreover, cyclic ordering of links in a vertex of a tree can also change, if μ crosses a singularity of the resultant variety (i.e. if $R(r_{kl}, r_{mn}) = 0$): the fat-graph structure of the forest is preserved only locally: under *small* variations of the 1-parametric families μ.

5.3 Map $f(x) = x^2 + c$. Julia sets, stability and preorbits

In this subsection we describe a few examples of orbits and pre-orbits to illustrate that normally the orbits tend to particular stable periodic orbits "in the future" and originate from entire set of unstable periodic orbits "in the past". We suggest to call the closure of this latter set $\partial J_A(f) = \overline{\mathcal{O}_-(f)}$ the boundary of algebraic Julia set of f. We demonstrate that there is no difference between the behaviour of *preorbits* of different points $x \in \mathbf{X}$: for x one can take invariant point (i.e. orbit of order one) of f, stable or unstable, a point of any periodic orbit or a point with unbounded grand orbit,– despite behaviour of the *orbits* is absolutely different in these cases. Often the boundary of Julia set is defined as pre-image of unstable invariant point, but our examples seem to demonstrate that this is unnecessary restriction.

Now, for given $c \in M(\mu)$ we can switch to what happens in the complex x plane (provided $\mathbf{X} = \mathbf{C}$). In this plane we have infinitely many periodic orbits. Their points form the set $\mathcal{O}(f_c) = \mathcal{O}_+(f_c) \cup \mathcal{O}_-(f_c)$, the union of all zeroes of all $F_n(x; f_c)$. There are a few stable periodic orbits and all the rest are unstable. For polynomial $f_c(x)$, all other stable orbits are actually located at $x = \infty$, unstable orbits separate them from the finite stable orbits. Generic grand orbit approaches either one of the stable orbits from $\mathcal{O}_+(f_c)$ or goes to infinity. The orbits which do *not* tend to infinity form **Julia set** $J(f_c) \subset \mathbf{X}$. The boundary of Julia set $\partial J(f_c)$ has at least three possible interpretations:

(i) it is a closure of the set of all unstable orbits, Fig. 5.3,

$$\partial J(f_c) = \overline{\mathcal{O}_-(f_c)};$$

(ii) it is a closure of a bounded grand orbit, associated with *any* single unstable periodic orbit (for example, with an orbit of order one, Fig. 5.4 – an unstable fixed point of f_c, – or any other, Fig. 5.6 and Fig. 5.7),

$$\partial J(f_c) = \overline{GO_1^-};$$

(iii) it is the set of end-points of all branches (not obligatory unstable, Fig. 5.8, or periodic, Fig. 5.9) of any generic grand orbit (this set is already full, no closure is needed).

Hypothetically all the three definitions describe the same object, moreover, for $f \in M$, i.e. when the map belongs to the Mandelbrot set, this object is indeed a boundary of some domain – the Julia set. If f quits the Mandelbrot set, the object is still well defined, but is not a boundary of anything: the "body" of Julia set disappears, only the "boundary" survives.

For the family $f_c = x^2 + c$ we know that there is exactly one stable orbit for all $c \in M(x^2 + c)$, see Fig. 5.2; for $c \notin M(x^2 + c)$ *all* orbits (not just some of them) go to $x = \infty$, so that $J(x^2 + c) = \emptyset$.

- $c \in M_1$. The set $\mathcal{O}_+$ consists of a single stable invariant (fixed) point (orbit of order one) $x_+^{(1)} = \frac{1}{2} - \frac{\sqrt{1-4c}}{2}$.
- $c \in M_2$. The set $\mathcal{O}_+$ consists of a single stable orbit of order two formed by the points $x_\pm^{(2)} = -\frac{1}{2} \pm \frac{\sqrt{-3-4c}}{2}$.

We now elaborate a little more on the definition (iii) of the Julia set boundary. Generic grand orbit is a tree with d links entering and one link exiting every vertex. Bounded (non-generic) grand orbit of order n ends in a loop, with $d - 1$ trees attached to every of the n vertices in the loop. Degenerate grand orbits (bounded or unbounded) have some branches eliminated, see the next subsection 5.4. In the rest of this subsection we consider non-degenerate grand orbits.

Going *backwards* along the tree we need to choose between d possible ways at every step. Particular path (branch) is therefore labeled by an infinite sequence of integers modulo d, i.e. by the sequence of elements of the ring $\mathbf{F_d}$ (if $d = p$ the ring is actually a field and the sequences are p-adic numbers). According to definition (iii) these sequences form the boundary of Julia set – this can be considered as a map between real line $\mathbf{R}$ (parameterized by the sequences) and ∂J_A, provided by a given grand orbit. Different grand orbits provide different maps. These maps deserve investigation. Here we just mention that among all the sequences the countable set of periodic ones is well distinguished, and they *presumably* are associated with periodic unstable orbits, which are dense in ∂J_A according to the definition (i).

Important example: The map $f(x) = x^2$ (the $c = 0$ case of $x^2 + c$) has a stable fixed point $x = 0$ and unstable orbits of order n consisting

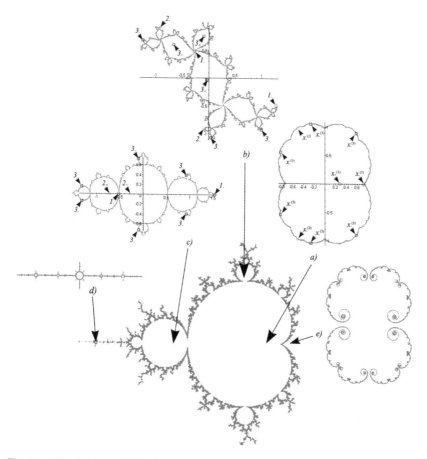

Fig. 5.3 The few lowest (of orders 1,2 and 3) periodic orbits. If all higher-order orbits are included, the closure provides the boundary ∂J of Julia set for the family $f_c(x) = x^2 + c$ by definition (i). The same orbits are shown on different fibers of Julia sheaf: a) $c = 0.2$, b) $c = -0.1 + 0.75i$, c) $c = -0.8$, d) $c = -1.76$, e) $c = 0.3$. For a)-c) values of $c \in M$ there is exactly one stable orbit *inside* the Julia set and all others are at the boundary.

of points on unit circle, which solve equation $x^{2^n} = x$ and do not lie on lower-order orbits:

$$\exp\left(2\pi i \frac{2^k + j}{2^n - 1}\right), \quad k = 0, \ldots, n - 1. \tag{5.7}$$

The Julia set – the attraction domain of $x = 0$ – is the interior of the unit circle (the unit disc) and all periodic unstable orbits fill densely its boundary, as requested by (i). Unstable fixed point is $x = 1$ and its grand

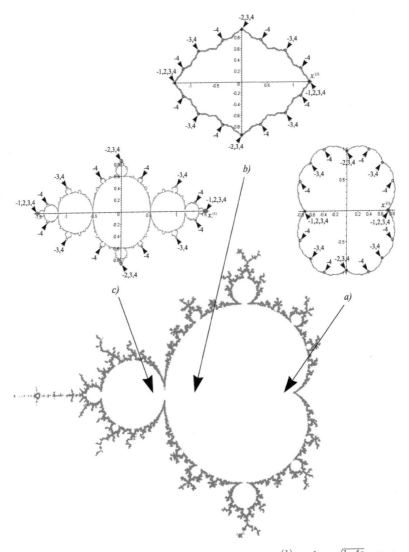

Fig. 5.4   Grand orbit (circles) of unstable fixed point $x_-^{(1)} = \frac{1}{2} + \frac{\sqrt{1-4c}}{2}$. Preimages of orders from 1 to 4 are shown. Its closure provides the boundary ∂J of Julia set for the family $f_c(x) = x^2 + c$ by definition (ii). The same grand orbit is shown for different fibers of Julia sheaf: a) $c = 0.2$, b) $c = -0.4$, c) $c = -0.8$. The grand orbit of unstable fixed point consists of all the "extreme" points ("spikes") of Julia set for Re $c < 0$, while these are the "pits" for $0 < $ Re $c < 1/4$. The pits are approaching the points of grand orbit of a stable fixed point, which lie inside the Julia set, and merge with them at $c = 1/4$, see Fig. 5.5. Both "spikes" and "pits" are dense everywhere in the boundary ∂J, if pre-images of all levels are included.

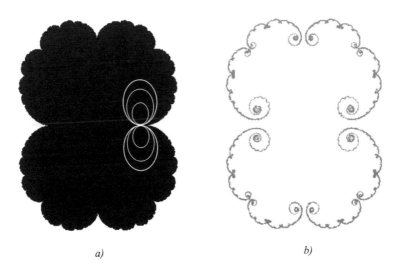

a)   b)

Fig. 5.5   Decomposition of Julia set at $c = 1/4$, where the two fixed points, stable and unstable,– and thus their entire grand orbits – coincide.   a) The pattern is shown as obtained by moving from *inside* the Mandelbrot set. White lines show peculiar circular f-flows in the vicinity of the stable fixed point (similar flow patterns occur in the vicinities of all other cusps). When the boundary of Mandelbrot set is crossed, all the cusps (located at all the points of the coincident grand orbits of two merged fixed points, which densely populate ∂J) dissociate into pairs of points, thus ∂J acquires holes (almost everywhere) and fails to be a boundary of anything: the "body" of Julia set "leaks away" through these holes and disappears. b) The resulting pattern as obtained by moving from *outside* the Mandelbrot set. Shown is unlarged piece of the Julia set, which now consists ony of the everywhere discontinuous boundary. The holes grow with increase of Re $c - \frac{1}{4} > 0$. The spiral structure inside the former Julia set is formed by separatrices between the flows, leaking through the holes and remaining "inside". Infinitely many separatrices (at infinitely many cusps) are formed by (infinitely many) branches of the grand orbit of the fixed point. At the same time each point on the grand orbit is a center of its own spiral. Spiral shape is related to the nearly circular shape of the f flows in the vicinity of the stable fixed point and its pre-images just before the formation of holes (for Re c slightly smaller than $\frac{1}{4}$). If the boundary of Mandelbrot set is crossed at another point, which is a zero of some other d_n (and belongs to some $\sigma_n^{(p)}$), the same role is played by grand orbits of the intersected orbits of order n: holes are formed at all points of these intersecting grand orbits.

orbit consists of all the points $\exp\left(\frac{2\pi i l}{2^m}\right)$, $l = 0, 1, \ldots, 2^m - 1$ of its m-th pre-images, which fill densely the unit circle, as requested by (ii). The grand orbit of any point $x = re^{i\varphi}$, which does no lie on a unit circle, $|x| = r \neq 1$, consists of the points x^{2^k}, with all integer k. For negative $k = -m$ the

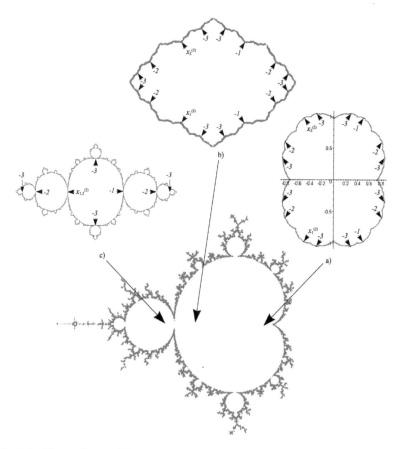

Fig. 5.6   The analogue of Fig. 5.4 for another grand orbit: the one of unstable orbit of order 2, formed by the points $x_{\pm}^{(2)} = -\frac{1}{2} \pm \frac{\sqrt{-3-4c}}{2}$. Again, the same grand orbit is shown for three different fibers of Julia sheaf: a) $c = 0.2$, b) $c = -0.4$, c) $c = -0.8$. This grand orbit also fills the boundary ∂J densely, in accordance with the definition (ii).

choice of one among 2^m phase factors $\exp\left(\frac{2\pi i l}{2^m}\right)$ should be made in order to specify one particular branch of the grand orbit. The branch can also be labelled by the sequence of angles $\{\varphi_m\}$, such that $\varphi_{m+1} = \frac{1}{2}(\varphi_m + 2\pi s_m)$, where $\{s_m\}$ is a sequence with entries 0 or 1. If the sequence is periodic with the period n, i.e. $s_{m+n} = s_m$, then $\varphi_{m+n} = \frac{1}{2^n}(\varphi_m + 2\pi p_m^{(n)})$, where $p_m^{(n)} = \sum_{i=0}^n 2^i s_{m+i}$. Then the origin of the corresponding branch is a periodic orbit of order n, containing the point on the unit circle with the phase $\varphi_m^{(n)} = \frac{2\pi p_n}{2^n - 1}$, which solves the equation $\varphi_m^{(n)} = \frac{1}{2^n}(\phi^{(n)} + 2\pi p_m^{(n)})$. Different

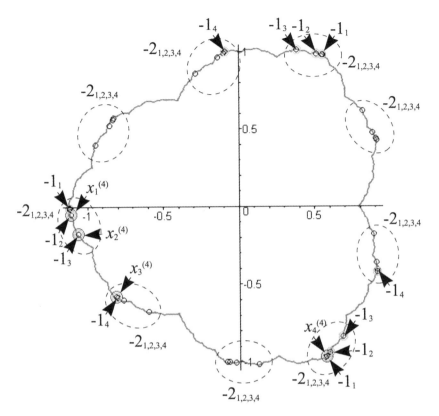

Fig. 5.7   Non-degenerate bounded grand orbit of order 4 for cubic $f(x)$, $d = 3$, $c = 0.3$ – one more illustration of the definition (ii). The first two preimages are shown. The dashed lines enclose each of the 9 second preimages of the whole orbit.

$m \in (m_0 + 1, \ldots, m_0 + n)$ describe n different elements of the periodic orbit, and m_0 labels the place where the sequence becomes periodic. For pure periodic sequences one can put $m_0 = 0$.

For $n = 1$ we have two options:

$$s_1 = 0, \quad p_1 = 0, \quad \varphi^{(1)} = 0 \quad \text{and}$$
$$s_1 = 1, \quad p_1 = 1, \quad \varphi^{(1)} = 2\pi \tag{5.8}$$

which have equivalent limits: the two branches originate at the single unstable orbit of order one, i.e. at the point $x = 1$.

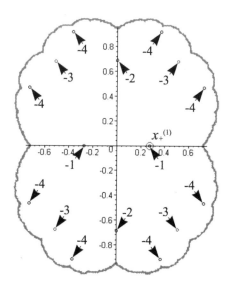

Fig. 5.8   Grand orbit of a stable fixed point $x_+^{(1)}$, $c = 0.2$. Its end- (or, better, starting-) points approach the boundary ∂J of Julia set, if one goes against the f flow and fill it densely, according to the definition (iii). The first four preimages are shown.

For $n = 2$ there are two options for the s-sequences:

$$s = \{0, 1\}, \quad p_2 = 2, \quad \varphi^{(2)} = \frac{4\pi}{3},$$

$$s = \{1, 0\}, \quad p_2 = 1, \quad \varphi^{(2)} = \frac{2\pi}{3}, \tag{5.9}$$

which describe the single existing orbit of order 2 as the origin for two different branches. The sequences $s = \{0, 0\}$ and $s = \{1, 1\}$ are actually of period one, not two.

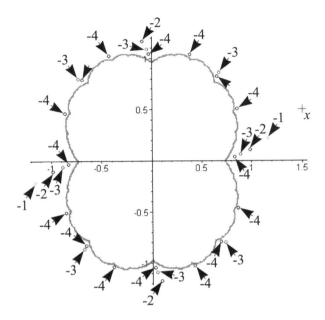

Fig. 5.9   Typical unbounded grand orbit at the same value of $c = 0.2$. Again, its origin fill densely the boundary of the same Julia set, ∂J, according to the definition (iii).

For $n = 3$ there are six options for the s-sequences:

$$s = \{0,0,1\}, \quad p_3 = 4, \quad \varphi^{(3)} = \frac{8\pi}{7},$$

$$s = \{0,1,0\}, \quad p_3 = 2, \quad \varphi^{(3)} = \frac{4\pi}{7},$$

$$s = \{0,1,1\}, \quad p_3 = 6, \quad \varphi^{(3)} = \frac{12\pi}{7},$$

$$s = \{1,0,0\}, \quad p_3 = 1, \quad \varphi^{(3)} = \frac{2\pi}{7},$$

$$s = \{1,0,1\}, \quad p_3 = 5, \quad \varphi^{(3)} = \frac{10\pi}{7},$$

$$s = \{1,1,0\}, \quad p_3 = 3, \quad \varphi^{(3)} = \frac{6\pi}{7}, \qquad (5.10)$$

which describe the two orbits, $(2,4,8) \times \frac{\pi}{7}$ and $(6,12,10) \times \frac{\pi}{7}$, of order 3 as the origins for six different branches.

Continuation to higher n is straightforward.

For more general map $f(x) = x^d$ we have

$$\varphi_{m+1} = \frac{1}{d}(\varphi_m + 2\pi s_m), \quad s_m \in (0, 1, \ldots, d-1) = G_d;$$

$$\varphi_m^{(n)} = \frac{2\pi p_n}{d^n - 1}, \quad p_n = \sum_{i=0}^{n-1} s_i d^i. \qquad (5.11)$$

In order to understand what happens when $c \neq 0$ we need to study the deformation of grand orbits with the variation of c. The deformation deforms the disc, but preserves the mutual positions of grand orbits on its boundary, i.e. the structure, partly shown in Fig. 5.10. As soon as we reach the boundary of the domain $M_1(\mu)$ in Mandelbrot set, the attractive fixed point loses stability and this means that its grand orbit approaches the boundary of the disc – and intersects there with an orbit of order n (if f approaches the point $M_1 \cap M_n \in \partial M_1$). Every point of GO_1 merges with n points of GO_n, which are located at different points of the disc boundary, so that the disc gets separated into sectors – the Julia set decomposes into a fractal-like structure, see Figs. 5.11 and 5.12. When f enters the domain M_n, the orbit O_n gets stable and its grand orbit fills densely the interior of the disc, while O_1 remains unstable and lies on the boundary, keeping it contracted (glued) at infinitely many points. Moreover, the angles α between the n components merging at contraction points increase smoothly from $\alpha = 0$ at $M_1 \cap M_n$ to $\alpha = \frac{2\pi}{n}$ when the next bifurcation point $M_n \cap M_l$ is reached. At $M_n \cap M_l$ new contractions are added – at the points, belonging to grand orbit GO_l, which become merging points of l/n sectors of our disc. And so on.

5.4 Map $f(x) = x^2 + c$. Bifurcations of Julia set and Mandelbrot sets, primary and secondary

The real slices of the lowest orbits and pre-orbits for the family $f_c(x) = x^2 + c$ are shown in Fig. 5.15. In order to understand what happens beyond real section, one can study Fig. 5.16.

In this case $z(c) = c$ – this is the point with degenerate pre-image, which consists of one rather than two points; this is ramification point of the Riemann surface of functional inverse of the analytic function $x^2 + c$. Actually, $z(c) = c$ for all the families $f_c(x) = x^d + c$, but then it has degeneracy $d - 1$. This $z(c)$ is f-image of the critical point $w(c) = 0$ of multiplicity $d - 1$.

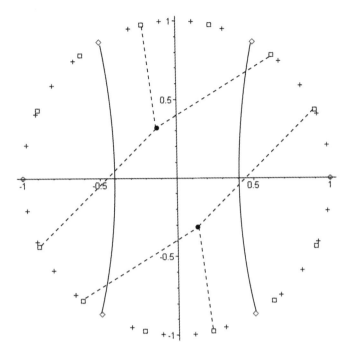

Fig. 5.10   Unit disc (Julia set at $c = 0$) with location of the first preimages (orbit and the first preorbit) of unstable orbits of orders one (circles), two (diamonds), three (boxes) and four (crosses). Locations of the points are shown for $c = 0$, but mutual positions of grand orbits remain the same for all complex c (though particular points can change order: a dot can pass through a cross – when $c = M_1 \cap M_2$, – but another dot will pass through the same cross in the opposite direction. The lines, connecting the two points of the orbit of order two and the two points of its first pre-image are also shown – they will be among the contracted separatrices in Fig. 5.11,– and analogous lines for an order-three orbit – to be contracted in Fig. 5.12.

For this family of maps one can straightforwardly study the pre-orbits and compare with the results of the resultant analysis – and this is actually done in Figs. 5.15, 5.16. Shown in these pictures are:

• the two fixed points (orbits of order one), i.e. zeroes of $G_1(x) = x^2 - x + c$:

$$x = \frac{1}{2} \pm \frac{\sqrt{1-4c}}{2};$$

Fig. 5.11   Julia set at $c = -\frac{3}{4}$, i.e. when $c = \sigma_1^{(1)} \cap \sigma_2^{(2)}$. See also Fig. 5.19.

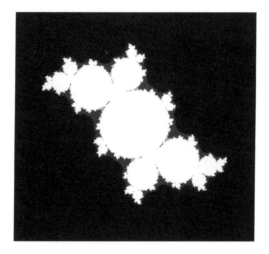

Fig. 5.12   Julia set at $c = -\frac{1}{8} + i\frac{3}{8}\sqrt{3}$, i.e. when $c = \sigma_1^{(1)} \cap \sigma_3^{(2)}$. See also Fig. 5.20.

- their first pre-images, i.e. zeroes of $G_{1,1}(x) = x^2 + x + c$:

$$x = -\frac{1}{2} \mp \frac{\sqrt{1-4c}}{2}$$

(the lower branch of $G_{1,1} = 0$ is mapped by f onto the upper branch of $G_1 = 0$, $(c,x) = (0,0)$ is a fixed point of a family $f_c(x)$);

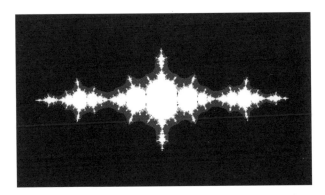

Fig. 5.13 Julia set at $c = -5/4$, i.e. when $c = \sigma_2^{(2)} \cap \sigma_4^{(3)}$. This time the pairs of points from the grand orbit of order two should be strapped, as well as the quadruples of points from a grand orbit of order four.

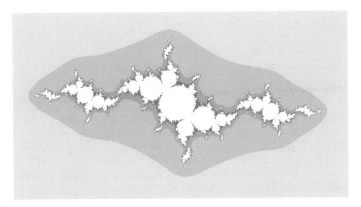

Fig. 5.14 Julia set at $c = -\frac{9}{8} + i\frac{\sqrt{3}}{8}$, i.e. when $c = \sigma_2^{(2)} \cap \sigma_6^{(3)}$. This time the pairs of points from a grand orbit of order two should be strapped, as well as the sextets of points from a grand orbit of order six.

- second pre-images, i.e. zeroes of $G_{1,2}(x) = x^4 + (2c+1)x^2 + c(c+2)$:

$$x = \pm\sqrt{-\frac{1+2c}{2} \mp \frac{\sqrt{1-4c}}{2}};$$

- the orbit of order two, i.e. zeroes of $G_2(x) = x^2 + x + c + 1$:

$$x = -\frac{1}{2} \pm \frac{\sqrt{-3-4c}}{2};$$

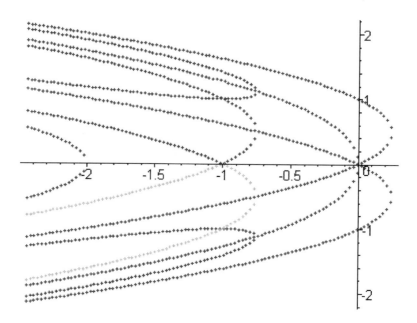

Fig. 5.15   Another type of section of Julia sheaf. The lowest two orbits of $f_c(x) = x^2 + c$: of orders one ($G_1(x;c) = 0$) and two ($G_2(x;c) = 0$) together with their first two pre-orbits ($G_{1,1}(x;c) = 0$, $G_{1,2}(x;c) = 0$ and $G_{2,1}(x;c) = 0$, $G_{2,2}(x;c) = 0$). The section Im $c = 0$, Im $x = 0$ of the entire 2-complex-dimensional pattern is shown, real c is on horizontal line, real x – on the vertical one. Distinguished points on the horizontal line, where the roots of various polynomials $G_{n,s}$ coincide, are: $c = 1/4$ – the zero of $d_1(c) = 1 - 4c$ and thus also of $D(G_{1,1}) \sim d_1$; $c = 0$ – the zero of $w_1(c) = c$ and thus also a zero of $W_{1,2}(c) = w_1(c)w_{1,2}(c)$ and of $D(G_{1,2}) \sim W_{1,2}$; $c = -3/4$ – the zero of $r_{12}(c) = 3 + 4c$, thus vanishing at this point are also discriminants $D(G_2)$, $D(G_{2,1})$, $D(G_{2,2})$ and resultants $r_{1,1|1,2}$, $r_{1,1|2,1}$, $r_{1,2|2,2}$, all proportional to r_{12}; $c = -1$ – the zero of $w_2(c) = c + 1$ and of the resultants $r_{2|2,1}$, $r_{2|2,2}$, $r_{2,1|2,2}$, all proportional to w_2; $c = -2$ – the second zero of $w_{1,2}(c) = c + 2$ and thus of $D(G_{1,2}) \sim W_{1,2}$. The zeroes of $w_{2,2} = c^2 + 1$, where the two branches of $G_{2,2} = 0$ intersect, are pure imaginary and not seen in this real section. $w_{1,1}$ and $w_{2,1}$ are identically unit for $d = 2$.

• its first pre-image, i.e. zeroes of $G_{2,1}(x) = x^2 - x + c + 1$:

$$\frac{1}{2} \pm \frac{\sqrt{-3 - 4c}}{2}.$$

(this time the lower branch of $G_{2,1} = 0$ is mapped by f onto the lower branch of $G_1 = 0$, $(c, x) = (-1, 0)$ is *not* a fixed point of a family $f_c(x)$);

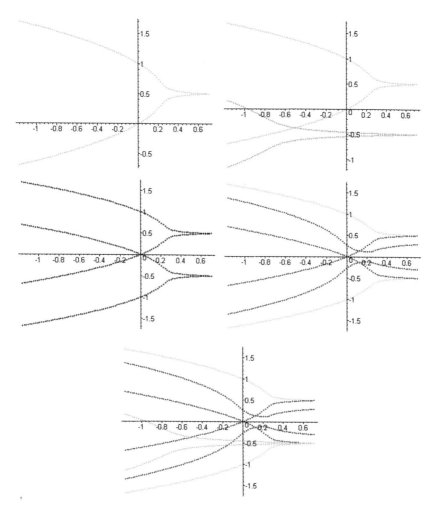

Fig. 5.16 Instead of the *section* of the two-complex-dimensional pattern by the plane Im $c = 0$, Im $x = 0$, shown in Fig. 5.15, this picture represents *projection* onto this plane, with Re c and Re x on the horizontal and vertical axes respectively. If it was projection of the same section, the only difference from Fig. 5.15 would be attaching of horizontal lines to the right of the bifurcation points, corresponding, say, to Re $\left(\frac{1}{2} \pm \frac{\sqrt{1-4c}}{2} \right) = \frac{1}{2}$ for $c > 1/4$ (while there are no traces of these complex roots in the *section*, they are well seen in *projection*). If we now slightly change the section, from arg$(c) = 0$ modπ to a small, but non-vanishing value (i.e. consider a section of Mandelbrot set not by a real axis, but by a ray, going under a small angle), then in projection we get the curves, shown in these pictures. a) The fixed points $x = \frac{1}{2} \pm \frac{\sqrt{1-4c}}{2}$. b) The same fixed points together with the orbit of order two, $x = -\frac{1}{2} \pm \frac{\sqrt{-3-4c}}{2}$. c) Fixed points together with their first pre-images, $x = -\frac{1}{2} \mp \frac{\sqrt{1-4c}}{2}$. d) The same with the second pre-orbit (zeroes of $G_{1,2}$) added. e) The same with the orbit of order two added.

- its second pre-images, i.e. zeroes of $G_{2,2}(x) = x^4 + (2c-1)x^2 + c^2 + 1$.

$$x = \pm\sqrt{\frac{1-2c}{2} \pm \frac{\sqrt{-3-4c}}{2}}.$$

From these pictures it is clear that along with the zeroes of ordinary discriminants and resultants (points $c = 1/4$ and $c = -3/4$), well reflected in the structure of universal Mandelbrot set, new characteristic points emerge: the zeroes of w_n and $w_{n,s}$ (like points $c = 0$, $c = -1$ and $c = -2$). Some of these points are shown in Fig. 5.17 together with the Mandelbrot set. By definition these additional points are associated with the boundary of Grand Mandelbrot set. It seems clear from Fig. 5.17 that there is also an intimate relation between some of these points with the *trail structure* of the Mandelbrot set itself. In Fig. 5.18 we show what happens to Julia set at zeroes of w_n and $w_{n,s}$. Clearly, zeroes of $w_{n,s}$ mark the places where Julia sets change the number of connected components.

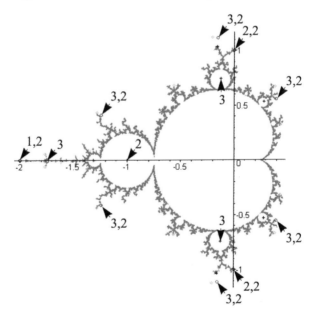

Fig. 5.17   Mandelbrot set, shown together with the zeroes of some w_n (crosses) and $w_{n,1}$ (circles), associated with the more general structure of Grand Mandelbrot set.

Discriminants for lowest pre-orbits are given by the following formulas (note that $d_2 = 1$ and all $w_{n,1} = 1$ for $d = 2$):

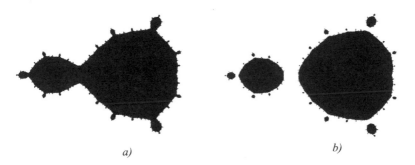

a)   b)

Fig. 5.18 This picture illustrates how the Julia set gets disconnected in the vicinity of zero $c = 4/9$ of $w_{1,1}$ for the (one-parametric) family $cx^3 + x^2$. Two views are shown: before and after the decay phase transition.

$n\backslash s$	$D(F_{ns})$	$D(G_{ns})$
11	$2^4 d_1^2 w_1^2$	d_1
12	$2^{16} d_1^4 w_1^7 w_{12}$	$2^4 d_1^2 w_1 w_{12}$
13	$2^{48} d_1^8 w_1^{18} w_{12}^3 w_{13}$	$2^{16} d_1^4 w_1^3 w_{12}^2 w_{13}$
14	$2^{128} d_1^{16} w_1^{41} w_{12}^7 w_{13}^3 w_{14}$	$2^{48} d_1^8 w_1^7 w_{12}^4 w_{13}^2 w_{14}$
15	$2^{320} d_1^{32} w_1^{88} w_{12}^{15} w_{13}^7 w_{14}^3 w_{15}$	$2^{128} d_1^{16} w_1^{15} w_{12}^8 w_{13}^4 w_{14}^2 w_{15}$
21	$2^8 d_1^2 r_{12}^6 w_1^2 w_2^2$	$-r_{12}$
22	$2^{32} d_1^4 r_{12}^{12} w_1^7 w_{12} w_2^6 w_{22}$	$2^4 r_{12}^2 w_{22}$
23	$2^{96} d_1^8 r_{12}^{24} w_1^{18} w_{12}^3 w_2^{15} w_{22}^3 w_{23}$	$2^{16} r_{12}^4 w_2^2 w_{22}^2 w_{23}$
24	$2^{256} d_1^{16} r_{12}^{48} w_1^{41} w_{12}^7 w_{13}^3 w_{14} w_2^{33} w_{22}^7 w_{23}^3 w_{24}$	$2^{48} r_{12}^8 w_2^2 w_{22}^4 w_{23}^2 w_{24}$
25	$2^{640} d_1^{32} r_{12}^{96} w_1^{88} w_{12}^{15} w_{13}^7 w_{14}^3 w_{15} w_2^{70} w_{22}^{15} w_{23}^7 w_{24}^3 w_{25}$	$2^{128} r_{12}^{16} w_2^5 w_{22}^8 w_{23}^4 w_{24}^2 w_{25}$
31	$2^{16} d_1^2 d_3^6 r_{13}^8 w_1^2 w_3^2$	$d_3^3 r_{13}^2$
32		$2^{12} d_3^6 r_{13}^4 w_{32}$
33		$2^{48} d_3^{12} r_{13}^8 w_{32}^2 w_{33}$
34		$2^{144} d_3^{24} r_{13}^{16} w_3 w_{32}^4 w_{33}^2 w_{34}$
35		$2^{384} d_3^{48} r_{13}^{32} w_3^2 w_{32}^8 w_{33}^4 w_{34}^2 w_{35}$
41		$d_4^4 r_{14}^3 r_{24}^2$
42		$2^{24} d_4^8 r_{14}^6 r_{24}^4 w_{42}$
43		$2^{96} d_4^{16} r_{14}^{12} r_{24}^8 w_{42}^2 w_{43}$
51		$d_5^5 r_{15}^4$
52		$2^{60} d_5^{10} r_{15}^8 w_{52}$
61		$d_6^6 r_{16}^5 r_{26}^4 r_{36}^3$

5.5   Conclusions about the structure of the "sheaf" of Julia sets over moduli space (of Julia sets and their dependence on the map f)

1. Julia set $J(f)$ is defined for a map $f_{\vec{c}} \in \mu \subset \mathcal{M}$. If $f \in \sigma_n(\mu)$, $J(f)$ is an attraction domain of the stable f-orbit of order n. Unstable periodic orbits and all bounded grand orbits (including pre-orbit of the stable orbit) lie at the boundary $\partial J(f)$. Every bounded grand orbit is dense in $\partial J(f)$. This property allows to define $\partial J(f)$ even for $f \notin M$, i.e. outside the Mandelbrot set: as a closure of any bounded grand orbit (or, alternatively, as that of a union of all periodic orbits). Just outside Mandelbrot set this closure fails to be a boundary of anything – there are no stable periodic orbits and Julia set $J(f)$ itself does not exist.

2. If $f \in \sigma_n(\mu) \subset M_1(\mu)$, the Julia set $J(f)$ is actually a deformation of unit disc. Its exact shape is defined by grand orbits of unstable periodic orbits, in particular all its "exterior points"/spikes are preimages of unstable fixed point(s). Its topology is even more transparent: dictated by position of f on the tree-powerful of the Mandelbrot set – as explained at the end of Sec. 5.3.

In more detail: $J(f)$ is a union of real domains in **X** (sectors of the unit disc):

$$J(f) = \cup_{\nu}^{\infty} J_{\nu}(f). \tag{5.12}$$

In variance with the Mandelbrot set decomposition into elementary domains σ_n's, the constituents J_{ν} (decomposition of the unit disc) and even their labelings ν are not uniquely defined.

2.1. At particular point (map) $f \in \sigma_k^{(p)} \cap \sigma_n^{(p+1)} = \partial\sigma_k^{(p)} \cap \partial\sigma_n^{(p+1)}$, where $R(G_k, G_n) = 0$ and two orbits O_k and O_n of orders $n = mk$ intersect, every point of O_k coincides with exactly $m = n/k$ merged points of O_n. Then every point of grand orbit GO_k is a singular point of $\partial J(f)$, where exactly m components from the set $\{J_{\nu}[O_n, O_k]\}$ "touch" together at zero angles, $\alpha = 0$, see examples of $(k, n) = (1, 2)$ in Fig. 5.11, $(k, n) = (1, 3)$ in Fig. 5.12, $(k, n) = (2, 4)$ in Fig. 5.13 and $(k, n) = (2, 6)$ in Fig. 5.14. There are no other singularities of $\partial J(f)$ at $f \in \sigma_k^{(p)} \cap \sigma_n^{(p+1)}$.

2.2. If we start moving f inside $\sigma_k^{(p)}$ ($k < n$), every m-ple from $\{J_{\nu}[O_n, O_k]\}$, which was *touching* at a single point for $f \in \sigma_k \cap \sigma_n$, acquires common boundary segments and the singularities get partly resolved: a point turns into m points and the full non-degenerate orbit O_n is restored, see Figs. 5.19 and 5.20. On this side of the (k,n)-"phase transition" the "order parameter" is the length of the common segments (separatrices): the deeper in σ_k the longer the segments. These segments can also be considered as peculiar unbounded grand orbit which densely fills "separatrices" between the merging orbits O_n (on $\partial J(f)$) and O_k (inside $J(f)$) when we approach the transition point from below – from the side of σ_k $(k < n)$.

2.3. If we start moving f inside $\sigma_n^{(p+1)}$ $(n > k)$, every m-ple from $\{J_\nu[O_n, O_k]\}$, which was *touching* at a single point for $f \in \sigma_k \cap \sigma_n$, continues to intersect at a single point – an orbit O_k is now unstable and should stay at the boundary $\partial J(f)$,– but angle α (the "order parameter" above the (k,n)-"phase transition") is no longer vanishing, but increases as we go deeper into σ_n, see Figs. 5.21 and 5.22. This is how the orbits O_k and O_n exchange stability: O_n leaves $\partial J(f)$ while O_k emerges at $\partial J(f)$. Since $n > k$ this is not so trivial to achieve and the problem is resolved by creating highly singular merging-points of m varieties at all the points of the grand orbit GO_k, when it emerges at $\partial J(f)$.

2.4. If, while moving inside $\sigma_n^{(p+1)}$ $(n > k)$, the map f reaches a new intersection point $\sigma_n^{(p+1)} \cap \sigma_N^{(p+2)}$, $N > n$, our (k,n)-structure (n/k-merged points at GO_k) feels this! Namely, when it happens, the angles $\alpha_{k,n}$ reach their maximal values of $\frac{2\pi}{m}$, our $m = n/k$ merging varieties turn into needles in the vicinity of the merging points (i.e. at points of GO_k), see Figs. 5.13 and 5.14.

2.5. All this describes the behaviour of J_ν sets, associated with the subvariety $\sigma_k^{(p)} \cap \sigma_n^{(p+1)} \subset \partial M(\mu)$ at the boundary between smooth constituents of Mandelbrot set. However, if we are at some point *inside* σ_k, there are many different structures of this type, coming from different boundary points (intersections with different σ_{km} with all integer m), and no one is distinguished. Thus the decomposition $J(f) = \cup_\nu J_\nu$ is ambiguous for f inside σ_k and gets really well defined only exactly at $\partial \sigma_k$ – though is an everywhere discontinuous function on this boundary, changes abruptly as f moves along $\partial \sigma_k$.

2.6. If f crosses the boundary $\partial M(\mu)$ and goes outside the Mandelbrot set then there are no stable orbits left and nothing can fill the disc: only the boundary remains, the Julia set has no "body" – it becomes a collection of disconnected curves, which are no longer boundaries of anything (but are still densely filled by bounded grand orbits and/or unstable periodic orbits). This dissociation of the boundary occurs at discriminant points, where the two orbits of the same order, one stable and another unstable on the Mandelbrot-set side cross and become unstable beyond the Mandelbrot set. When they diverge after crossing, the orbits (and their entire grand orbits) pull away their pieces of the boundary of Julia set, and the "body" of Julia set "leaks away" through emerging holes, see Fig. 5.5.

Despite it is not exhaustive, this decomposition structure carries significant information, in fact it is information about *topology* of Julia sets and singularity structure of their boundaries. A remote region $\sigma_n^{(p)}$ in the main component of Mandelbrot set $M_1(\mu)$ can be reached from $\sigma_1^{(1)}$ by a path which lies entirely inside $M(\mu)$ and passes through a sequence of points $\sigma_k \cap \sigma_l$ connecting different components. Passage through every such separation point changes Julia set in the above-described way, drastically changes its topology and angles α. However, the change of the *shape* (i.e. geometry rather than just topology) of $J(f)$ when f is varied *inside* every particular component σ_n is not described in equally exhaustive manner.

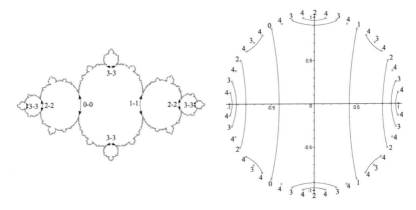

Fig. 5.19   The deformation of Fig. 5.11: Julia set in the vicinity of intersection point $c = \sigma_1^{(1)} \cap \sigma_2^{(2)}$ with $c \in \sigma_1^{(1)}$. It is obtained from the unit disk by strapping the *pairs* of points on the boundary, which belong to grand orbit of the orbit of order 2. Domains J_ν overlap and separatrices connect the points of grand orbit of order 2, which are strapped at the intersection (phase transition) point. The length of separatrices can serve as an order parameter in this phase ($\sigma_1^{(1)}$). Actually, there are infinitely many different order parameters inside $\sigma_1^{(1)}$, associated with transitions to different $\sigma_n^{(2)}$. Separatrices are closures of peculiar (not-bounded) f-orbit and of its pre-images.

In other words, the suggested theory gives a complete description of the following characteristics of Julia set:
   – "Extreme" points (like "spikes" and "pits" in Fig. 5.4) – they belong

to grand orbits of unstable fixed points (orbits of order one);

– Merging points of sectors and needles – they belong to grand orbits, which became unstable on the path of f from inside the region $\sigma_1^{(1)}(\mu)$ in Mandelbrot set $M_1(\mu)$, where the order-one orbit was stable. Every merging point is associated to particular "bridge crossing" at the point $\sigma_n \cap \sigma_k$ in $M_1(\mu)$, $k|n$, where the stable orbit O_k "exchanges stability" with O_n. The number of merging components is equal to $m = n/k$. The angles between merging components of Julia set depend on how far we are from the corresponding "bridge crossing", and change from $\alpha = 0$ (touching) at the bridge to $\alpha = \frac{2\pi}{m}$ (needle) when the next bridge is reached.

3. The shape of $\partial J(f)$ is defined pure algebraically, because all the bounded grand orbits lie densely inside $\partial J(f)$. However, in variance with the boundary ∂M of Mandelbrot sets, $\partial J(f)$ cannot be decomposed into smooth constituents (like $\partial \sigma_n$ which are smooth almost everywhere): no countable set of points from bounded grand orbits fill densely any smooth variety (like the zeroes of $R(G_k, G_{km})$ did for given k and arbitrary m in the case of $\partial \sigma_k$).

4. Unbounded (generic) grand orbits lie entirely either inside $J(f)$ or beyond it. Each unbounded grand orbit originates in the vicinity of entire $\partial J(f)$: for every point of $\partial J(f)$ and for every grand orbit there is a branch which comes (with period d) close enough to this point. If grand orbit lies in $J(f)$, it tends to a stable orbit – of order n if $f \in \sigma_n$. If grand orbit lies outside of $J(f)$, it tends to infinity. One can say that infinity is a "reservoir" of stable orbits: if the degree d of polynomial map f is increased, the new stable orbits come from infinity.

5. For $f \in M_{n\alpha}$ lying in other components of Mandelbrot set, disconnected from M_1 (but linked to it by densely populated *trails*), the structure of Julia set is similar, with the only exception: the starting point is not a single disc, but a collection of discs, formed when the map f leaves M_1 and steps onto one or another trail leading to the other components $M_{n\alpha}$. Both periodic orbits and pre-orbits jump between the boundaries (if unstable) or interiors (if stable) of these discs. With this modification, the same description of Julia sheaf is valid, with "monodromies" described by the same procedure of interchanging stable grand orbits inside with unstable grand orbits on the boundary, which causes fractalization of the boundary.

6. If $f \notin M(\mu)$ gets outside the Mandelbrot set, the Julia set is no longer a disc or collection of discs: it looses its "body" (which would be occupied by the stable bounded grand orbit) and consists only of the boundary (where unstable periodic orbits and their grand orbits live). Its bifurcations can still be described algebraically, but now the intersections of unstable orbits will be the only ones to play the role. These bifurcations are controlled by the Grand Mandelbrot set.

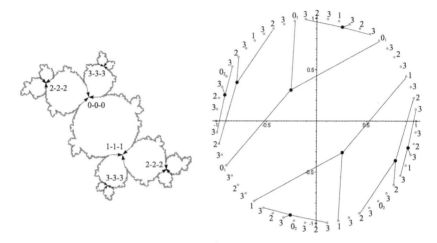

Fig. 5.20   Analogous deformation of Fig. 5.12: Julia set in the vicinity of intersection point $c = \sigma_1^{(1)} \cap \sigma_3^{(2)}$ with $c \in \sigma_1^{(1)}$. It is obtained by strapping the *triples* of points on the boundary of the disk, belonging to a grand orbit of the order 3.

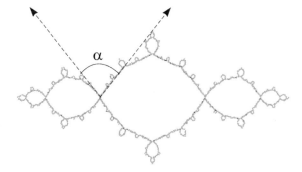

Fig. 5.21 Deformation of Fig. 5.11 in another direction: Julia set in the vicinity of intersection point $c = \sigma_1^{(1)} \cap \sigma_2^{(2)}$ with $c \in \sigma_2^{(2)}$. In this phase the order parameter is the angle $\alpha > 0$ between the merging domains. It increases from $\alpha = 0$ at transition point $c = \sigma_1^{(1)} \cap \sigma_2^{(2)}$ to the maximal value $\alpha = 2\pi/m = \pi$ when c reaches other points of the boundary $\partial \sigma_2^{(2)}$, i.e. when the next phase transition occurs, e.g. when $c = \sigma_2^{(2)} \cap \sigma_4^{(3)}$ (Fig. 5.13) or $c = \sigma_2^{(2)} \cap \sigma_6^{(3)}$ (Fig. 5.14).

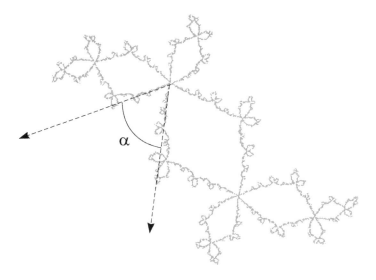

Fig. 5.22 Analogous deformation of Fig. 5.12: Julia set in the vicinity of intersection point $c = \sigma_1^{(1)} \cap \sigma_3^{(2)}$ with $c \in \sigma_3^{(2)}$. In this phase the angles α change from $\alpha = 0$ at $c = \sigma_1^{(1)} \cap \sigma_3^{(2)}$ to $\alpha = 2\pi/m = 2\pi/3$ at any other point of the boundary $\partial \sigma_3^{(2)}$.

Chapter 6

Other examples

In this section we present some more examples of the few-parametric families of maps $f_c(x)$ in order to show that our considerations and conclusions are in no way restricted to the peculiar case of $f_c(x) = x^2 + c$. For every family $\mu \subset \mathcal{M}$ we describe the Mandelbrot set $M(\mu)$ as the union of stability domains $M_n(\mu)$, its boundary $\partial M(\mu)$ as the closure of appropriate discriminant variety, the structure of grand orbits and of algebraic Julia sets.

We are going to illustrate and support the following claims:

• Consideration of various 1-parametric families shows universality of the tree structure of Mandelbrot sets and non-universality of particular numbers: shows the difference between n, k, m- and α-parameters.

• Particular 1-parametric families (like $\{f_c(x) = x^d + c\}$ or $\{f(x) = cx^d + x^2\}$) are not in generic position – even among the maps of given degree – and the corresponding sections of Mandelbrot/discriminant variety are not fully representative: have additional special features, symmetry properties and accidental degeneracies.

• 2-parametric families can be considered as families of 1-parametric families. Their analysis helps to understand how the non-universal components of the description change and helps to understand their nature and relation to next-level algebraic structures like singularities of resultant varieties, controlled by the resultants of the higher order. Clearly, non-universality is nothing but the corollary of our restriction to ordinary resultants and further work will provide unique interpretation to all properties of particular sections/families μ.

In variance with the previous chapter 5 we do not give exhaustive treatment of every particular example below. Each example serves to illustrate one or another particular aspect of the problem and a particular line of further development.

6.1 Equivalent maps

Different maps f and $\tilde{f}$ can be equivalent from the point of view of our considerations if they are related by a diffeomorphism ϕ of $\mathbf{X}$, $f \circ \phi = \phi \circ \tilde{f}$, or

$$f(\phi(x)) = \phi(\tilde{f}(x)). \tag{6.1}$$

Then, obviously, $f^{\circ n} \circ \phi = \phi \circ \tilde{f}^{\circ n}$, $F_n \circ \phi = \phi \circ \tilde{F}_n$, $G_n \circ \phi = \phi \circ \tilde{G}_n$ and so on. If f and $\tilde{f}$ are polynomials of the same degree – what will be the most important application of equivalence below – than ϕ should be a linear transformation, $\phi(x) = ax + b$.

6.2 Linear maps

6.2.1 *The family of maps $f_{\alpha\beta} = \alpha + \beta x$*

In this case

$$F_n(x; f_{\alpha\beta}) = (\beta^n - 1)\left(x + \frac{\alpha}{\beta - 1}\right)$$

and decomposition formula (4.1) reads:

$$F_n = \prod_{k|n}^{\tau(n)} G_k,$$

$$G_1(x; f_{\alpha\beta}) = (\beta - 1)x + \alpha,$$

$$G_k(x; f_{\alpha\beta}) = g_k(\beta), \quad k > 1. \tag{6.2}$$

The functions $g_k(\beta)$ are circular polynomials:

$$\beta^n - 1 = \prod_{k|n}^{\tau(n)} g_k(\beta)$$

with

$$g_1(\beta) = \beta - 1,$$
$$g_2(\beta) = \beta + 1 = -g_1(-\beta),$$
$$g_3(\beta) = \beta^2 + \beta + 1,$$
$$g_4(\beta) = \beta^2 + 1,$$
$$g_5(\beta) = \beta^4 + \beta^3 + \beta^2 + \beta + 1,$$
$$g_6(\beta) = \beta^2 - \beta + 1 = g_3(-\beta),$$
$$\dots \qquad (6.3)$$

For simple $p > 2$

$$g_p(\beta) = \frac{\beta^p - 1}{\beta - 1} = \beta^{p-1} + \dots + \beta + 1,$$

$$g_{2p}(\beta) = \frac{\beta^p + 1}{\beta + 1} = \beta^{p-1} - \beta^{p-2} + \dots - \beta + 1,$$

$$g_{p^k}(\beta) = \frac{\beta^{p^k} - 1}{\beta^{p^{k-1}} - 1} = \beta^{p(p-1)} + \dots + \beta^p + 1$$

and so on.

All $G_k(x; f_{\alpha\beta})$ in (6.2) except for G_1 do not depend on x. This means that there are no orbits of order $k > 1$ unless β is appropriate root of unity – when there are infinitely many such orbits. For example, when $\beta = -1$ any pair $(x, \alpha - x)$ form a second-order orbit of the map $x \to \alpha - x$; and when $\beta = \pm i$ any quadruple $(x, \pm ix + \alpha, -x + (1 \pm i)\alpha, -ix \pm i\alpha)$ is an orbit of order four etc.

Of course, the family of linear maps is a highly degenerate example, still this peculiar section of Mandelbrot/discriminant variety should be reproduced in particular limits of more general families of maps below.

6.2.2 *Multidimensional case*

We use the chance to illustrate the new ingredients of the theory which arise in multidimensional situation.

Take $\mathbf{X} = \mathbf{C^r}$ and consider the linear map $f : x \to Bx$ where B is an $r \times r$ matrix. The iterated map remains linear, $f^{\circ n}(x) = B^n x$ and $F_n(x) = (B^n - I)x$. Irreducible components $G_k = g_k(B)$ are independent of x for $k > 1$, just as in the one-dimensional situation. However, for $r > 1$ degeneration pattern is richer: of interest are all situations when two eigenvalues coincide, matrices acquire Jordan form and some two eigenvectors become collinear.

The case of $r = 2$. The eigenvector $x = (x_1, x_2)$ of $B = \begin{pmatrix} b_{11} & b_{12} \\ b_{21} & b_{22} \end{pmatrix}$ is a quadric in $\mathbf{CP^1}$ in homogeneous coordinates, i.e. satisfies

$$\frac{1}{2}x_i Q_{ij}(B) x_j = b_{12}x_2^2 + (b_{11} - b_{22})x_1 x_2 - b_{21}x_1^2 = 0. \qquad (6.4)$$

The linear map B is degenerate when determinant of the matrix $Q(B)$ (or, what is the same, discriminant of the polynomial $x_i Q_{ij} x_j / x_2^2$ in x_1/x_2 vanishes),

$$\det_{2 \times 2} Q(B) = (b_{11} - b_{22})^2 + 4b_{12}b_{21} = 0.$$

We call this expression discriminant of the linear map B. For $r > 2$ consideration is a little more sophisticated and we do not discuss it here, see [11]. The resultant of two maps A and B vanishes when some eigenvector of A gets collinear to some eigenvector of B. Of course, different iterations of the same linear map have the same eigenvectors, and resultant analysis for linear maps remain trivial. Still, it is important to remember about these structures in analysis of non-linear multidimensional maps.

6.3 Quadratic maps

This class of examples helps to demonstrate, that even equivalent maps (related by diffeomorphism in $\mathcal{M}$) can have differently *looking* Mandelbrot sets. Of course, this is not a big surprise: non-linear change of variables (say, $c \to c^2$) in $\mathcal{M} \cap \mu$ changes the shape and even the number of domains in Fig. 5.2.

6.3.1 *Diffeomorphic maps*

Consideration of generic 3-parametric family of quadratic maps $f_{\alpha\beta\gamma}(x) = \alpha + \beta x + \gamma x^2$ can actually be reduced to that of $f_c(x) = x^2 + c$ by the rule (6.1), because

$$\gamma f_{\alpha\beta\gamma}(x) + \frac{\beta}{2} = f_c(\gamma x + \frac{\beta}{2})$$

provided

$$c = \alpha\gamma + \frac{\beta}{2} - \frac{\beta^2}{4}. \qquad (6.5)$$

Then $\gamma f_{\alpha\beta\gamma}^{\circ 2}(x) + \frac{\beta}{2} = f_c(\gamma f_{\alpha\beta\gamma}(x) + \frac{\beta}{2}) = f_c(f_c(\gamma x + \frac{\beta}{2})) = f_c^{\circ 2}(\gamma x + \frac{\beta}{2})$
and, in general,

$$\gamma f_{\alpha\beta\gamma}^{\circ n}(x) + \frac{\beta}{2} = f_c^{\circ n}(\gamma x + \frac{\beta}{2}).$$

In other words, $f_{\alpha\beta\gamma} : \mathcal{M} \to \mathcal{M}$ is diffeomorphic to $f_c : \mathcal{M} \to \mathcal{M}$. Discriminants and resultants for $f_{\alpha\beta\gamma}$ can be obtained from those of f_c by a substitution (6.5). However, since (6.5) is non-linear (quadratic) transformation, it can change the numbers of zeroes, and the numbers of elementary domains in sections of $\mathcal{D}^*$ can be different for different 1-parametric subfamilies of $\{f_{\alpha\beta\gamma}\}$.

For two 2-parametric families to be analyzed below we have equivalences:

$$x^2 + px + q : \quad p = \beta, \ q = \alpha\gamma, \ c = \frac{4q + 2p - p^2}{4}$$

and

$$\gamma x^2 + (b+1)x : \quad p = b+1, q = 0, \ c = \frac{1 - b^2}{4}.$$

6.3.2 Map $f = x^2 + c$

First of all, we summarize the discussion of Mandelbrot set in the previous chapter 5 in the form of a table:

$$r_{17} = (4c)^6 + (4c)^5 + (4c)^4 + (4c)^3 + 15 \cdot (4c)^2 - 17 \cdot (4c) + 127,$$

$$r_{18} = 256c^4 + 32c^2 - 64c + 17,$$

and so on. Stars in the table stand for too long expressions. $\mathcal{N}_n^{(p)}$ is the number of domains $\sigma_n^{(p)}$ of given level p and order n, which still differ by the values of parameters m_i and α_i that are ignored in this table. For one-parametric families the number q of cusps often depends only on the level p. Taking this into account, we can write down – and check in this and other examples – the set of *sum rules* for the numbers $\mathcal{N}_n^{(p)}$:

$$\sum_{p=0}^{\infty} \mathcal{N}_n^{(p)} = \deg_c[G_n(w_c, c)],$$

$$\sum_{p=0}^{\infty} q(p)\mathcal{N}_n^{(p)} = \deg_c[d_n],$$

$$\sum_{p=1}^{\infty} \mathcal{N}_n^{(p)} = \sum_m \deg_c[r_{n,n/m}]. \tag{6.6}$$

$f = x^2 + c$						
n	1	2	3	4	5	6
# of orbits $= \deg_x[G_n]/n$	2	1	2	3	6	9
# of el. domains $= \deg_c[G_n(w_c, c)]$	1	1	3	6	15	27
d_n total # of cusps $= \deg_c[d_n]$	1-4c	i	-7-4c	$64c^3 + 144c^2 + 108c + 135$	see eq.(5.5)	see eq.(5.6)
	1	0	1	3	11	20
$r_{n,n/m}$ and total # of (n,n/m) touching points = $\deg_c[r_{n,n/m}]$ for						
m=2	-	r_{12} $3+4c$	-	r_{24} $4c+5$	-	r_{36} $64c^3 + 128c^2 + 72c + 81$
	-	1	-	1	-	3
m=3	-	-	r_{13} $16c^2 + 4c + 7$	-	-	r_{26} $16c^2 + 36c + 21$
	-	-	2	-	-	2
m=4	-	-	-	r_{14} $16c^2 - 8c + 5$	-	-
	-	-	-	2	-	-
m=5	-	-	-	-	r_{15} $256c^4 + 64c^3 + 16c^2 - 36c + 31$	-
	-	-	-	-	4	-
m=6	-	-	-	-	-	r_{16} $16c^2 - 12c + 3$
	-	-	-	-	-	2
$\mathcal{N}_n^{(p)}$: # of el. domains $\sigma_n^{(p)} \subset S_n$ with $q(p)$ cusps						
$p = 0, \ q = 1$	1	0	1	3	11	20
$p = 1, \ q = 0$	0	1	2	2	4	3
$p = 2, \ q = 0$	0	0	0	1	0	4
$p = 3, \ q = 0$	0	0	0	0	0	0
. . .						

Actually, for every given n only finitely many values of p contribute: p is the number of non-unit links in the branch of the multipliers tree, thus $p \leq \log_2 n$. From (6.6) one can deduce an identity which does not include sums over p:

$$\mathcal{N}_n^{(0)} + \sum_m \deg_c[r_{n,n/m}] = \deg_c[G_n(w_c, c)], \qquad (6.7)$$

and in many cases (including the families $x^d + c$) the r.h.s. is just $\mathcal{N}_n(d)$. Also in many cases (again including $x^d + c$) $q(0) = d - 1$ and $q(p) = d - 2$

for $p > 0$. In such situations

$$(d - 1)\mathcal{N}_n^{(0)} + (d - 2) \sum_m \deg_c[r_{n,n/m}] = \deg_c[d_n], \qquad (6.8)$$

and we also have a consistency condition between (6.7) and (6.8):

$$\deg_c[d_n] + \sum_m \deg_c[r_{n,n/m}] = (d - 1)\deg_c[G_n(w_c, c)]. \qquad (6.9)$$

Coming back to our particular family $x^2 + c$, one can easily check that consistency condition is satisfied by the data in the table, and also $\mathcal{N}_n^{(0)} = \deg_c[d_n]$, as requested by (6.8).

For analysis of Julia sheaf the above table should be supplemented by the data concerning pre-orbits and associated resultants, but this remains beyond the scope of the present book.

6.3.3 *Map $f_{\gamma\beta0} = \gamma x^2 + \beta x = \gamma x^2 + (b + 1)x$*

According to (6.5) we have

$$c = \frac{\beta}{2} - \frac{\beta^2}{4} = \frac{1}{4} - \frac{(\beta - 1)^2}{4} = \frac{1}{4}(1 - b^2), \qquad (6.10)$$

and one can expect that quadratic cusp for domain σ_1 at $c = \frac{1}{4}$ disappears after such transformation. In fact, transformation (6.10) makes from cardioid σ_1 in c-plane a pair of unit discs in b-plane with centers at $b = 0$ and $b = 2$, which touch each other at the point $b = 1$, $S_1 = \sigma^{(0)}[1+] \cup \sigma^{(0)}[1-]$: this is a good example, demonstrating that links with $m = 1$ in multiples tree sometimes contribute to the structure of Mandelbrot set.

The boundary ∂S_1 is described by a system of equations (4.19):

$$\partial S_1 : \quad \begin{cases} |2\gamma x + \beta| = 1 \\ \gamma x^2 + \beta x = x \end{cases}$$

i.e. $x = 0$ or $x = \frac{1-\beta}{\gamma}$, $|\beta| = 1$ or $|\beta - 2| = 1$ and $|1 \pm b| = 1$. Accordingly the resultants $r_{1n}(b)$ should have roots lying on these two unit circles. Actually, they are all made from products of circular polynomials (6.3):

$$r_{1n}(b) = \gamma^{N_n} g_n(\beta) g_n(2 - \beta) = \gamma^{N_n} g_n(1 + b) g_n(1 - b),$$

$$r_{2,2n} = \gamma^{N_{2n}} g_n(5 - b^2),$$

$$\dots$$

in particular (for $\gamma = 1$)

$$\gamma^{-2} r_{12}(b) = 3 + 4c = (\beta + 1)(3 - \beta) = (2 + b)(2 - b),$$

$$\gamma^{-6} r_{13}(b) = 16c^2 + 4c + 7 = (\beta^2 + \beta + 1)(\beta^2 - 5\beta + 7)$$

$$= (3 + 3b + b^2)(3 - 3b + b^2),$$

$$\gamma^{-12} r_{14}(b) = 16c^2 - 8c + 5 = (\beta^2 + 1)(\beta^2 - 4\beta + 5)$$

$$= (2 + 2b + b^2)(2 - 2b + b^2),$$

$$\gamma^{-30} r_{15}(b) = 256c^4 + 64c^3 + 16c^2 - 36c + 31$$

$$= (\beta^4 + \beta^3 + \beta^2 + \beta + 1)(\beta^4 - 9\beta^3 + 31\beta^2 - 49\beta + 31)$$

$$= (5 + 10b + 10b^2 + 5b^3 + b^4)(5 - 10b + 10b^2 - 5b^3 + b^4),$$

$$\gamma^{-54} r_{16}(b) = 16c^2 - 12c + 3 = (\beta^2 - \beta + 1)(\beta^2 - 3\beta + 3)$$

$$= (1 + b + b^2)(1 - b + b^2),$$

$$\ldots$$

$$\gamma^{-12} r_{24}(b) = 4c + 5 = 6 - b^2 = 1 + (5 - b^2),$$

$$\gamma^{-54} r_{26}(b) = 16c^2 + 36c + 21 = b^4 - 11b^2 + 21$$

$$= (5 - b^2)^2 + (5 - b^2) + 1,$$

$$\ldots$$

$$\gamma^{-108} r_{36}(b) = 64c^3 + 128c^2 + 72c + 81 = -b^6 + 11b^4 - 37b^2 + 108,$$

$$\ldots$$

The last expression, for r_{36}, has nothing to do with circular polynomials: the map (6.10) transforms $\sigma_1^{(0)}$, which is quadratic cardioid in c-plane, into a bouquet of two unit discs in b-plane; it converts $\sigma_2^{(1)}$, which is a disc in c-plane, into domain $|b^2 - 5| < 1$ in b-plane; but it does not simplify higher σ_n, which are complicated in c-plane and remain complicated in b-plane.

What is interesting in this example, dependence on γ is trivial in b-plane, and the pattern of Mandelbrot set (its section by quadratic maps) remain the same in the limit of $\gamma \to 0$. In this limit the biggest part of Mandelbrot set is associated with orbits, lying at very big $x \sim \gamma^{-1}$. Naive limit $\gamma = 0$, considered in the above Sec. 6.2, ignores such orbits and only the unit disc $|\beta| < 1$ (domain $\sigma_1^{(0)}$) is seen by examination of linear maps. However, the other parts are actually present (nothing happens to them in the limit of $\gamma = 0$), just not revealed by consideration of strictly linear maps. Similarly, consideration of quadratic maps alone can ignore (overlook) other pieces of Mandelbrot set, and so does every restriction to maps of a given degree.

Powers of γ in the above formulas for resultants characterize intersection of orbits at infinity. Since in this particular case all periodic orbits except for a single fixed point tend to $x = \infty$ as $\gamma \to 0$, the asymptotic of resultants is defined by the following rule (see Eq. (6.13) below):

$$R(G_k, G_l) = r_{kl}^l \sim \gamma^{N_k(N_l - \delta_{l,1})}, \quad k > l,$$

where $N_l = N_l(d = 2)$ is the degree of the polynomial $G_l(x; f)$ for quadratic map f, and N_l/l is the number of periodic orbits of order l.

We omit the table for this family, because all the numbers in it are obtained by doubling the corresponding numbers in the table for $x^2 + c$ (provided degrees of all polynomials are counted in terms of b rather than c).

6.3.4   *Generic quadratic map and* $f = x^2 + px + q$

Expressions for generic quadratic map $\gamma x^2 + \beta x + \alpha$ can be obtained from formulas below by a trivial change of variables: $x \to \gamma x$, $p = \beta$, $q = \alpha\gamma$.

$$G_1 = F_1 = x^2 + (p - 1)x + q,$$

$$G_2 = \frac{F_2}{G_1} = x^2 + (p + 1)x + (p + q + 1),$$

$$\cdots$$

$$G_{1,1} = \frac{F_{1,1}}{G_1} = G_2 - 1 = x^2 + (p + 1)x + (p + q + 1),$$

$$G_{1,2} = \frac{F_{1,2}}{G_1 G_{1,1}} = x^4 + 2px^3 + (2q + p^2 + p + 1)x^2$$

$$+ (2pq + p^2 + p)x + (q^2 + pq + 2q + p),$$

$$G_{2,1} = \frac{F_{2,1}}{G_1 G_2 G_{1,1}} = x^2 + (p - 1)x + (q + 1),$$

$$G_{2,2} = \frac{F_{2,2}}{G_1 G_2 G_{1,1} G_{1,2} G_{2,1}}$$

$$= x^4 + 2px^3 + (p^2 + 2q + p - 1)x^2 + (p^2 + 2pq - p)x + (q^2 + pq + 1)$$

$$\cdots$$

The critical point is $w = -p/2$, the "irreversible point" $z = f(w) = \frac{4q - p^2}{4q}$. Accordingly, for generic map $w = -\beta/2\gamma$ and $z = \frac{4\alpha\gamma - \beta^2}{4\gamma}$.

$$w_1 = q + \frac{p}{2} - \frac{p^2}{4},$$

$$w_2 = 1 + w_1,$$

$$w_3 = 1 + 2pq + q + q^3 + \frac{p}{2} + \frac{p^2}{4} - \frac{3p^2q^2}{4} + \frac{3pq^2}{2} -$$

$$-\frac{p^2q}{2} - \frac{3qp^3}{4} - \frac{p^4}{16} - \frac{3p^3}{8} - \frac{p^6}{64} + \frac{3p^4q}{16} + \frac{3p^5}{32},$$

$$\dots$$

$$w_{11} = 1,$$

$$w_{12} = 2 + w_1,$$

$$\dots$$

6.3.5   *Families as sections*

We now use these examples to briefly discuss the geometric interpretation of the theory. The Mandelbrot set in Fig. 5.2 is obtained as a union of infinitely many elementary domains σ in the plane of complex c, glued together in a very special way along the discriminant variety (which has *real* codimension one and is a closure of a union of complex varieties $\mathcal{R}^*_{k,n}$ of *complex* codimension one). Each domain σ can be described as a real domain, defined by some equation of the form $\Sigma(c) < 1$ with a real function $\Sigma(c)$ of a complex variable c.

Switching to the two-parametric family $f = x^2 + px + q$, we get the Mandelbrot set, which is a similar union of domains $\tilde{\sigma}$ in the space $\mathbf{C^2}$ of two complex variables p and q, with each $\tilde{\sigma}$ defined by the equation $\tilde{\Sigma}(p,q) < 1$, where essentially $\tilde{\Sigma}(p,q) = \Sigma\left(c = q + \frac{1}{4} - \frac{1}{4}(p-1)^2\right)$.

Since we cannot draw pictures in $\mathbf{C^2}$, we restrict the whole pattern to real p and q: this badly spoils the nice picture, but preserves the main property which we are going to discuss. After restriction to real c, Fig. 5.2 turns into collection of segments (implicit in the lower part of that figure), see Fig. 6.1. Instead we can now extend the picture in another direction: to two-dimensional plane (p,q). Segments turn into parabolic strips, see Fig. 6.2, where different families, like $x^2 + c$ and $x^2 + (b+1)x$, are represented as different sections. It is clear from the picture why one and the same Mandelbrot set looks very different if restricted to different families. Another version of the same pattern is presented in Fig. 6.3, where Mandelbrot consists of straight rather than parabolic strips (after complexification strips become "cylindrical" or, better, toric domains). However, beyond quadratic family such ideal "torization" is not possible, different resultant varieties cannot be made exactly "parallel", because sometime they intersect, see Eqs. (6.12) for the simplest example – at level of cubic families. The best one can try to achieve is the representation in terms of somewhere-strapped tori, represented by chains of sausages, see Fig. 7.1 at the end of the book. It is an interesting question, whether Mandelbrot set can be represented as a toric variety.

6.4   Cubic maps

Generic cubic map – the 4-parametric family $f_{\alpha\beta\gamma\delta}(x) = \alpha + \beta x + \gamma x^2 + \delta x^3$ – is diffeomorphic to a 2-parametric family $f_{p,q}(x) = x^3 + px + q$,

$$p = \beta - \frac{\gamma^2}{3\delta},$$

$$q = \frac{2\gamma^3 + 9(1 - \beta)\gamma\delta + 27\alpha\delta^2}{27\delta\sqrt{\delta}}.$$

In addition to some initial information about this 2-parametric family we provide a little more details about several 1-parametric sub-families of cubic maps,

$$f_c(x) = x^3 + c \quad \text{with} \quad (p, q) = (0, c),$$

$$f_c(x) = cx^3 + x^2 \quad \text{with} \quad (p, q) = \left(-\frac{1}{3c}, \frac{9c + 2}{27c\sqrt{c}}\right),$$

$$f_c(x) = ax^3 + (1 - a)x^2 + c \quad \text{with} \quad (p, q) = \left(-\frac{(1 - a)^2}{3a}, \frac{2(1 - a)^3 + 27a^2c}{27a\sqrt{a}}\right)$$

respectively (in the last case a is considered as an additional parameter).

Our main purpose will be to demonstrate four (inter-related) new phenomena, not seen at the level of quadratic maps:

– Keeping degree of the maps fixed, we ignore (hide) the biggest part of Julia set, which is detached from the "visible" part and is located at $x = \infty$. Julia set decays at peculiar "decay" bifurcation points, which are presumably zeroes of $w_{n,s}(f)$. See Fig. 5.18.

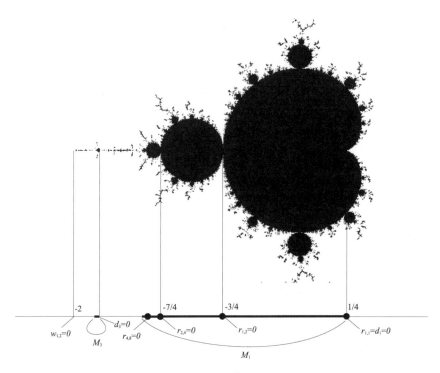

Fig. 6.1   The Mandelbrot set for the family $x^2 + c$ shown together with its section by a real line Im $c = 0$, where it turns into collection of segments. These segments form disconnected groups – sections of different components M_n (actually, only the big group, associated with M_1, and a tiny one, associated with M_3, are seen in the picture; if the section was not by a real line, a single M_1 would provide many disconnected groups of segments). Within each group the segments, corresponding to different $\sigma_k^{(p)}$ touch by their ends, real zeroes of resultants, and free ends are real zeroes of discriminants (perhaps, those of pre-orbit polynomials, i.e. zeroes of d_n or $w_{n,s}$).

– For essentially multi-parametric families μ of maps (like the 2-parametric $f_{p,q}(x) = x^3 + px + q$), different components r_{kn} and d_n of the resultant/discriminant variety intersect and higher order singularities of this variety can be revealed in this way. At these singularities particular sections of universal Mandelbrot set, associated with 1-parametric families, reshuffle, and new components $M_{n\alpha}$ can split from those previously in existence. Thus these reshufflings are responsible for formation of the trail structure of particular Mandelbrot sets.

– If the $\mathbf{Z_d}$ symmetry of the map $f(x) = x^d + c$ is slightly, but fully broken (e.g. by addition of a term βx or γx^2 with small non-vanishing β

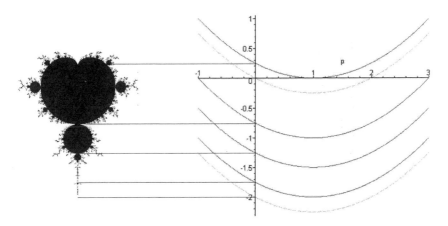

Fig. 6.2 The same picture in the *real* (p,q) plane, p is plotted along the abscissa axis, q – along ordinate. This section of Mandelbrot set represents it as a collection of domains bounded by parallel parabolas, some domain touch (have common boundaries, which consist of zeroes of the resultants – now they are not points, but codimension-one parabolas).

or γ), then the $\mathbf{Z_{d-1}}$ symmetry of the section of Mandelbrot set gets broken in peculiar way: the central domain M_1 and the nearest (in a relevant topology, actually, those with largest sizes ρ_n) components $M_{n\alpha}$ continue to possess the $\mathbf{Z_{d-1}}$ symmetric shape, while remote components $M_{n\alpha}$ acquire the non-symmetric form (characteristic for the Mandelbrot set for the family $x^2 + c$). See Fig. 6.4. Detailed analysis of reshuffling process, shown in Figs. 6.5 and 6.6, reveals how the non-central components and trail structure of Mandelbrot set is formed.

 – In generic situation a given map (at a given point in the Mandelbrot space) can possess *several* stable periodic orbits (not just one as it happens in the case of x^2+c – and this brings us closer to continuous situation, where several stable points, limiting cycles or other attractors can co-exist). Bifurcations of Julia set occur whenever any of these orbits exchange stability with some unstable one. See Fig. 6.7.

6.4.1 *Map* $f_{p,q}(x) = x^3 + px + q$

$$G_1(x) = x^3 + (p-1)x + q,$$

$$G_2(x) = x^6 + (2p+1)x^4 + 2qx^3 + (p^2+p+1)x^2 + ((2p+1)q-1)x + p + q^2 + 1,$$

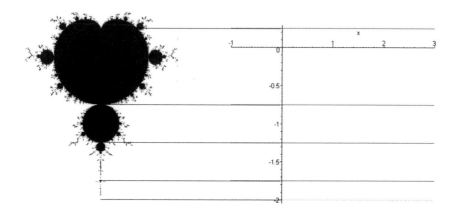

Fig. 6.3   The same picture in the real (c, p)-plane: p plotted on abscissa axis, c – on ordinate. The Mandelbrot set is represented by collection of straight horizontal strips. Resultants and discriminants depend only on c and not on p thus their zeroes form vertical lines, separating the strips. In general situation (beyond quadratic families) the Mandelbrot set cannot be made "cylindrical": the obstacle is provided by intersections of different resultant and discriminant varieties, controlled by higher resultants and discriminants, see (6.12) for examples. The best one can hope to achieve is a "sausage-chain" representation, see Fig. 7.1 at the end of this book.

$$G_{1,1}(x) = G_2 - 1.$$

The two critical points are $w = \pm\sqrt{-p/3}$, or $p = -3w^2$. Here are some irreducible components of pre-orbit resultants and discriminants (they are all products of two factors, associated with $+w$, and $-w$, such products are polynomials in integer powers of p; we list expressions which are not too long):

$$
\begin{aligned}
w_1 &= (q - 2w^3 - w)(q + 2w^3 + w) = \frac{1}{27}(27q^2 + 4p^3 - 12p^2 + 9p), \\
w_2 &= (4w^6 - 2w^4 - 4w^3q - 2w^2 + wq + q^2 + 1) \\
&\quad \times (4w^6 - 2w^4 + 4w^3q - 2w^2 - wq + q^2 + 1), \\
w_{1,1} &= (q - 2w^3 + 2w)(q + 2w^3 - 2w) = \frac{1}{27}(27q^2 + 4p^3 + 24p^2 + 36p), \\
w_{2,1} &= (q^2 + 1 - 2wq + w^2 - 4w^3q + 4w^4 + 4w^6) \\
&\quad \times (q^2 + 1 + 2wq + w^2 + 4w^3q + 4w^4 + 4w^6), \\
&\quad \cdots
\end{aligned}
$$

Some discriminants and resultants:

$$
\begin{aligned}
d_1 &= -(27q^2 + 4p^3 - 12p^2 + 12p - 4), \\
r_{12} &= 27q^2 + 4p^3 - 12p^2 + 16,
\end{aligned}
$$

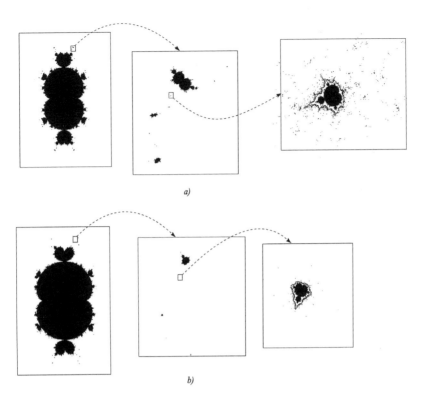

Fig. 6.4 Different, close and remote, components of the Mandelbrot set in the complex c plane for the family $ax^3 + (1 - a)x + c$ for different values of a (considered as an additional parameter): a) $a = 4/5$, b) $a = 2/3$. One can see how the symmetry of different domains is changing with the variation of a: the smaller a the larger is the low-symmetry domain. See Fig. 6.5 for description of transition (reshuffling) process and Sec. 6.4.5 for comments. Halo around the domain in the right pictuires (at high resolution) is an artefact and should be ignored.

$$d_2 = 27q^2 + 4p^3 + 24p^2 + 48p + 32,$$

$$r_{1|1,1} = 27(q + w + 2w^3)(q - w - 2w^3) = 27w_1,$$

$$d_{1,1} = \frac{D(F_{1,1})}{d_1^3 r_{1|1,1}^2} = \frac{D(G_{1,1})}{d_1^2}$$

$$= -27(q - 2w + 2w^3)(q + 2w - 2w^3) = -27w_{1,1},$$

$$r_{1|1,2} = 27(q + w + 2w^3)(q - w - 2w^3) = 27w_1,$$

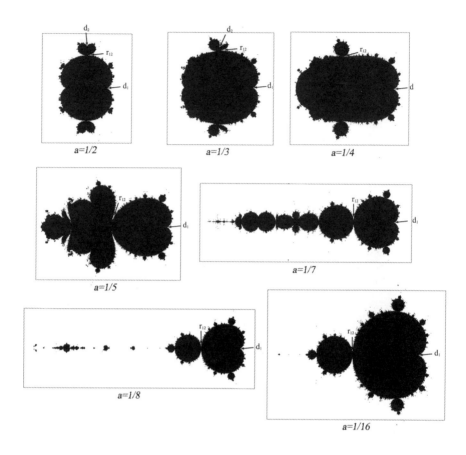

Fig. 6.5   Shown is a sequence of views of Mandelbrot sets for different values of a in the 2-parametric family $\{ax^3 + (1 - a)x^2 + c\}$, interpolating between the 1-parametric families $\{x^3 + c\}$ at $a = 1$ and $\{x^2 + c\}$ at $a = 0$. See Sec. 6.4.5 for additional comments and footnote 2 in that section for **precaution** concerning this particular figure: only the right parts of the pictures can be trusted, the left parts should be reflections of the right. Arrows show positions of some zeroes of d_1, d_2 and r_{12}. The full sets of these zeroes are shown in fully reliable (though less picturesque) Fig. 6.6.

$$r_{1,1|1,2} = R(G_{1,1}, G_{1,2})^{1/2} = 27(q + w + 2w^3)(q - w - 2w^3) = 27w_1,$$

$$\cdots$$

Zeroes of $w_{1,1}(f)$ form a submanifold, where the Julia set can decay into two disconnected components $((p, q) = (-3/4, 3/4)$ is a particular point on

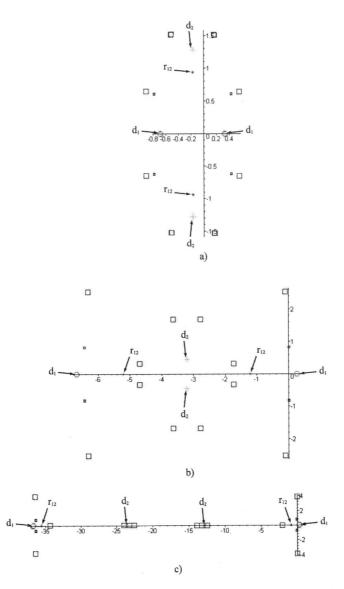

Fig. 6.6   A sequence of pictures, showing positions in the complex c-plane of zeroes of $d_1(c)$ (big circles), $d_2(c)$ (big crosses), $d_3(c)$ (big boxes), $r_{12}(c)$ (small crosses), $r_{13}(c)$ (small boxes) for the 2-parametric family $\{ax^3 + (1-a)x^2 + c\}$, interpolating between the 1-parametric families $\{x^3 + c\}$ at $a = 1$ and $\{x^2 + c\}$ at $a = 0$: a) $a = 2/3$, b) $a = 1/6$, c) $a = 1/15$. These are particular special points in the patterns, presented in Fig. 6.5. Three arrows point to a zero of d_1, a zero of r_{12} and a zero of d_2. After the two roots of r_{12} merge at the real-c line, this happens at $a = \frac{5-\sqrt{21}}{2} \approx 1/5$, they start moving in opposite directions along this line. We arbitrarily choose to point at the right of the two zeroes (because the left one is not seen in the partly erroneous Fig. 6.5).

this submanifold, another point is $(p, q) = (0, 0)$, where Julia set turns into an ideal circle). As cubic f degenerates into quadratic map, namely, when $q \to 0$ with p fixed, one of these two components travels to $x = \infty$. See Fig. 5.18.

Since in this subsection we are dealing with a two-parametric family, we can observe intersections of various components $r_{k,n}$ and d_n of the universal discriminant variety (these intersections lie in complex codimension two). They are seen in the section by complex-dimension-two family $f = x^3 + px + q$ as particular points: zeroes of the resultants of $d(p, q)$ and $r(p, q)$, considered as polynomials of q for a fixed p or vice versa. For example:

$$R(r_{12}, d_1|q) = 16 \cdot 3^6 (3p - 5)^2,$$
$$R(r_{12}, d_2|q) = 16 \cdot 3^6 (3p + 2)^2,$$
$$R(d_1, d_2|q) = 16 \cdot 3^{10} (p^2 + p + 1)^2;$$

$$(6.11)$$

$$R(r_{12}, d_1|p) = -2^{10} (729q^2 + 32),$$
$$R(r_{12}, d_2|p) = 2^8 (729q^2 + 256),$$
$$R(d_1, d_2|p) = 2^8 3^9 (27q^4 + 16).$$

6.4.2   Map $f_c = x^3 + c$

$$G_1 = x^3 - x + c,$$

$$G_2 = x^6 + x^4 + 2cx^3 + x^2 + cx + c^2 + 1,$$

$$G_{1,1} = G_2 - 1,$$

$$\cdots$$

The double stable point is $w = 0$, and

$$w_1 = c^2,$$

$$w_2 = (1 + c^2)^2,$$

$$w_3 = (1 + c^2 + 3c^4 + 3c^6 + c^8)^2,$$

$$w_4 = c^8,$$

$$w_{11} = c^2,$$

$$w_{12} = c^2(c^4 + 3c^2 + 3)^2,$$

$$\cdots$$

$$w_{21} = (1 + c^2)^2,$$

$$w_{22} = 1 + 3c^4 + 2c^6 + 3c^8 + 3c^{10} + c^{12},$$

$$\cdots$$

$$w_{31} = (1 + c^2 + 3c^4 + 3c^6 + c^8)^2,$$

$$\cdots$$

These quantities are obviously factorizable – in seeming contradiction with our claim that they are all irreducible. In fact, factorization is accidental and is lifted by infinitesimal deformation, for example, by taking $p \neq 0$, see the previous subsection 6.4.1. Such deformation also breaks the accidental coincidences, like that of w_2 and w_{21}.

$$d_3 = 3^{18}c^{12} + 2 \cdot 3^{19}c^{10} + 3^{13}3571c^8 + 2^4 3^9 11 \cdot 19 \cdot 107c^6$$
$$+ 2^5 3^8 13^2 137c^4 + 2^6 3^4 13^4 17c^2 + 2^8 13^{16}.$$

$$d_4 = (150094635296999121c^{24} + 1200757082375992968c^{22}$$
$$+ 4203267461712259335c^{20} + 8399740516065395253c^{18}$$
$$+ 10909964351274746418c^{16} + 10526401881511556976c^{14}$$
$$+ 8522156414444085612c^{12} + 5544611719418268000c^{10}$$
$$+ 2750472027922567500c^8 + 1314354779366400000c^6$$
$$+ 4599012556800000000c^4 + 1677721600000000000) \cdot$$
$$\cdot (282429536481c^{16} + 1757339338104c^{14} + 4642459719687c^{12}$$
$$+ 6806074010589c^{10} + 6891783220746c^8 + 5994132959232c^6$$
$$+ 4118269132800c^4 + 1739461754880c^2 + 1073741824000) =$$
$$= 3^{36}c^{24} + \ldots = (27c^2)^{12} + \ldots$$

d_4 appears factorizable, but this is an accidental factorization, lifted by any deformation of the symmetric family $x^3 + c$.

The sum rules (6.6) now imply:

$$\mathcal{N}_n^{(0)} + \sum_m \deg_c[r_{n,n/m}] = \deg_c[G_n(w_c, c)],$$

$f = x^3 + c$	n	1	2	3	4	5
# of orbits $= \deg_x[G_n]/n$		3	3	8	18	48
# of el. domains $= \deg_c[G_n(w_c,c)]$		1	2	8	24	80
d_n		$4 - 27c^2$	$(27c^2+32)i$	*	*	?
total # of cusps $= \deg_c[d_n]$		2	2	12	40	152
$r_{n,n}/m$ and total # of $(n,n/m)$ touching points $= \deg_c[r_{n,n/m}]$ for						
m=2		–	$\overset{r_{12}}{\dfrac{27c^2+16}{2}}$	–	$\overset{r_{24}}{\dfrac{729c^4 + 1620c^2 + 1000}{4}}$	–
m=3		–	–	$\overset{r_{13}}{\dfrac{729c^4 + 27c^2 + 169}{4}}$	–	–
m=4		–	–	–	$\overset{r_{14}}{\dfrac{729c^4 - 324c^2 + 100}{4}}$	–
m=5		–	–	–	–	$\overset{r_{15}}{\dfrac{3^{12}c^8 + 3^9 4c^6 + 3^6 46c^4 - 7\cdot 27\cdot 263c^2 + 11^4}{8}}$
...						
$\mathcal{N}_n^{(p)}$: # of el. domains $\sigma_n^{(p)} \subset S_n$ with $q(p)$ cusps						
p=0, q=2		1	0	4	16	72
p=1, q=1		0	2	4	4	8
p=2, q=1		0	0	0	4	0
p=3, q=1		0	0	0	0	0
....						

$$2\mathcal{N}_n^{(0)} + \sum_m \deg_c[r_{n,n/m}] = \deg_c[d_n],$$

$$\sum_{p \geq 1} \mathcal{N}_n^{(p)} = \sum_m \deg_c[r_{n,n/m}]$$

and the consistency condition

$$\deg_c[d_n] + \sum_m \deg_c[r_{n,n/m}] = 2\deg_c[G_n(w_c, c)],$$

all obviously satisfied by the data in the table (question marks near some numbers in the table indicate that they were found not independently, but with the help of the sum rules).

With these explicit expressions we can illustrate the small general theorem, formulated and proved after Eq. (4.22). Namely, solutions to the system

$$\begin{cases} G'_n/H_n = 0 \\ G_n = 0 \end{cases}$$

where $H_n = \{F'_n, G_n\}$, are given by zeroes of *irreducible* discriminant d_n.

For $n = 1$ we have: $G'_1 = 3x^2 - 1$, the system has solutions, i.e. G'_1 and G_1 possess a common zero, when their resultant $D(G_1) = 4 - 27c^2$ vanishes, and $H_1 = -6x$ never has common zeroes with G'_1. Thus solutions of the system are zeroes of the irreducible discriminant d_1, which in this case coincides with $D(G_1)$.

For $n = 2$ $G'_2 = 6x^5 + 4x^3 + 6cx^2 + 2x + c$ and $H_2 = -36x(x^3 + x + c)(x^3 + c)(x^3 - x + c)$ both depend on c and have a common root whenever their resultant $R(H_2, G'_2) = -2^{10}3^{13}c^4(27c^2 - 16)(27c^2 + 16)(27c^2 + 8)$ vanishes. At the same time, the system can have solution only when G'_2 and G_2 possess common zeroes, i.e. when $D(G_2) = (27c^2 + 16)(27c^2 + 32)^2$ vanishes. We see that a pair of the roots, $27c^2 + 16$ is actually eliminated, because they are also zeroes of $R(H_2, G'_2)$ and what remains are the roots of the irreducible discriminant $d_2 = 27c^2 + 32$.

The first few pre-orbit discriminants for the family $x^3 + c$ are:

$$ns \qquad\qquad D(G_{ns})$$

$$11 \qquad\qquad -3^3 d_1^2 w_{11}$$
$$12 \qquad\qquad -3^{27} d_1^6 w_1 w_{11}^3 w_{12}$$
$$13 \qquad\quad -3^{135} d_1^{18} w_1^4 w_{11}^9 w_{12}^3 w_{13}$$
$$14 \quad -3^{567} d_1^{54} w_1^{13} w_{11}^{27} w_{12}^9 w_{13}^3 w_{14}$$

$$21 \qquad\qquad 3^6 d_2^4 r_{12}^2 w_{21}$$
$$22 \qquad\qquad 3^{54} d_2^{12} r_{12}^6 w_{21}^3 w_{22}$$

$$31 \qquad\qquad 3^{24} d_3^6 r_{13}^4 w_{31}$$

The Julia sheaf for this family is shown in Fig. 6.8.

6.4.3   Map $f_c(x) = cx^3 + x^2$

$$G_1 = cx^3 + x^2 - x,$$

$$G_2 = c^3 x^6 + 2c^2 x^5 + c(c+1)x^4 + 2cx^3 + (c+1)x^2 + x + 1,$$

$$G_{1,1} = G_2 - 1,$$

$$\cdots$$

Critical points $w = \{0, -\frac{2}{3c}\}$. Contribution of critical point $w = 0$ can be ignored in most applications, and we define in this case:

$$w_1 = G_1(w) = F_1(w) = \frac{2}{3c^2}\left(c + \frac{2}{9}\right),$$

$$F_2(w) = \frac{2}{3c^5}\left(c + \frac{2}{9}\right)\left(c^3 - \frac{2}{9}c^2 + \left(\frac{2}{9}\right)^2 c + 2\left(\frac{2}{9}\right)^3\right),$$

$$F_{1,1}(w) = F_2(w) - F_1(w) = -\frac{4}{27c^5}\left(c - \frac{4}{9}\right)\left(c + \frac{2}{9}\right)^2,$$

$$G_{1,1}(w) = \frac{F_{1,1}(w)}{F_1(w)} = -\frac{2}{9c^3}\left(c - \frac{4}{9}\right)\left(c + \frac{2}{9}\right),$$

$$w_{1,1} = \frac{G_{1,1}(w)}{G_1(w)} = -\frac{1}{3c}\left(c - \frac{4}{9}\right).$$

$c = 4/9$ means that $(p, q) = (-3/4, 3/4)$.

$f = cx^3 + x^2$				
n	1	2	3	4
# of orbits $= \deg_x[G_n]/n$	3	3	8	18
# of el. domains $= \deg_c[G_n(\varepsilon,c)]$	1	3	12	36
d_n total # of cusps $= \deg_c[d_n]$	$4c+1$ 1	$ic^5/2\,(4-13c+32c^2)$ 2	* 10	* 31
$r_{n,n/m}$ and total # of $(n,n/m)$ touching points $= \deg_c[r_{n,n/m}]$ for m=2	- -	$\overset{r12}{c^5(6c+3)}$ 1	- -	$\overset{r24}{c^{105}(100c^3-220c^2+47c+20)}$ 3
m=3	- -	- -	$\overset{r13}{c^{22}(169c^2+68c+7)}$ 2	- -
m=4	- -	- -	- -	$\overset{r14}{c^{70}(100c^2+44c+5)}$ 2
$\mathcal{N}_n^{(p)}$: # of el. domains $\sigma_n^{(p)} \subset S_n$ with $q(p)$ cusps p=0, q=1	1	2?	10?	31?
p=1, q=0	0	?	?	?
p=2, q=0	0	?	?	?
p=3, q=0	0	?	?	?
...				

$$d_3 = c^{66}(1235663104c^{1}0 - 765891776c^9 + 315356704c^8$$
$$-107832976c^7 + 33146817c^6 - 9493768c^5 + 2585040c^4$$
$$+233040c^3 - 123072c^2 + 42752c + 7168)$$
$$= c^{66}(2^8 13^6 c^{10} + \ldots)$$

$$d_4 = c^{500}(+18014398509481984000000000000000c^{31} - 39631676720860364800000000000000c^{30}$$
$$+8693002811987722240000000000000c^{29} - 1304466312653910835200000000000c^{28}$$
$$+117375780269187085107200000000c^{27} - 2515218215953699438592000000c^{26}$$
$$+571799816500523292950528000c^{25} + 3182928112090865666668016640c^{24}$$
$$+44985877595412915098027744c^{23} - 22393560229455792230880276c^{22}$$
$$-4844558197092071004917119c^{21} - 391359614494908199135640c^{20}$$
$$+847227219438292501416048c^{19} + 319824545156577533637200c^{18}$$
$$+43548434837473605814592c^{17} + 33835232982782581502464c^{16}$$
$$+6794824688429177434112c^{15} - 295875264228762906624c^{14}$$
$$-183114267457753161728c^{13} + 165660797805589659648c^{12} + 93048634733415038976c^{11}$$
$$+19011408301729447936c^{10}$$
$$+25910347327455363072c^9 + 16153160956010037248c^8 + 4359984836267474944c^7$$
$$+933600163325280256c^6 + 314124845396787200c^5 + 94391968968212480c^4$$
$$+17132514753642496c^3 + 1782088231550976c^2 + 99381248262144c + 2319282339840)$$
$$= c^{500}(2^{67} 5^{13} c^{31} + \ldots),$$

$$r_{15} = c^{236}(14641c^4 + 12506c^3 + 4021c^2 + 576c + 31) = c^{236}((11c)^4 + \ldots)$$
$$r_{23} = c^{72},$$
$$r_{25} = c^{720},$$
$$r_{34} = c^{864},$$
$$\ldots$$

It is instructive to reproduce the powers of c in these expressions by direct analysis of roots of the polynomials $G_n(x)$ to demonstrate the strong correlation between different roots of iterated polynomials. We consider here only the simplest case: the periodic orbits of orders one and two in the limit $c \to 0$. In this limit all polynomials $G_n(x; f)$ for cubic map f should turn into the same polynomials, but for quadratic map f, thus the powers of the polynomials should decrease appropriately: from $N_n(d = 3)$ to $N_n(d = 2)$. In particular, the degrees of $G_1(x)$ and $G_2(x)$ should change from 3 to 2 and from 6 to 2 respectively. For cubic map $G_1(x; f)$ has three roots, one of them grows as $\frac{1}{c}$, the other two remain finite, so that $G_1(x) \sim c\left(x + \frac{1}{c}\right)x(x-1) + O(c)$. (In this case the roots can be found exactly, they are: $x = 0$ and $x = \frac{1}{2c}(-1 \pm \sqrt{1 + 4c})$.) $G_2(x)$ has six roots, four grow as $c \to 0$, and two remain finite. However, the four roots cannot all grow as $\frac{1}{c}$, because then $G_2(x) = c^3 \prod_{i=1}^{6}(x - \rho_i^{(2)})$ would grow as $c^{3-4} = c^{-1}$ instead of turning into finite quadratic polynomial $x^2 + x + 1 + O(c)$. Actually, only two of the four routs grow as $\frac{1}{c}$, while the other two – only as $\frac{1}{\sqrt{c}}$.

Still, this is only the beginning. Unless the coefficients in front of the singular terms are carefully adjusted, we will not reproduce correct c-asymptotics of d_1, d_2 and r_{12}. Actually, the first discriminant is simple:

$$d_1 = D(G_1) \sim c^{2 \cdot 3 - 2}\left(\frac{1}{c}\right)^{2 \cdot 2} \sim c^0 + O(c),$$

where the first factor is the standard α_d^{2d-2} from discriminant definition, and the second comes from squared differences between the singular root of $G_1(x)$ and two finite ones.

Similarly we can naively estimate:

$$r_{12} \sim c^{3 \cdot 3 + 1 \cdot 6}\left(\frac{\cdots}{c} - \frac{\cdots}{c}\right)^2 \left(\frac{\cdots}{c} - \frac{\cdots}{\sqrt{c}}\right)^2 \left(\frac{\cdots}{c} - \rho^{(2)}\right)^2 \left(\rho^{(1)} - \frac{\cdots}{c}\right)^4$$

$$\cdot \left(\rho^{(1)} - \frac{\cdots}{\sqrt{c}}\right)^4 \left(\rho^{(1)} - \rho^{(2)}\right)^4 \overset{?}{\sim} c^{15-2-2-2-4-2}(1 + O(c)) \sim c^3(1 + O(c)),$$

where the first factor is $\alpha_d^{\frac{1}{2}((N_1-1)N_2 + N_1 N_2)}$ (see Sec. 6.7), the three next come from differences between the singular root of G_1 and the roots of G_2, while the remaining three terms come from the differences of two finite roots of G_1 with the roots of G_2 (here $\rho^{(1)}$ and $\rho^{(2)}$ denote *finite* roots of G_1 and G_2, there are two and two). However, here we run into a problem: the asymptotics c^3 is wrong, actually $r_{12} \sim c^5(1 + O(c))$. The reason for this is that the factor $\left(\frac{\cdots}{c} - \frac{\cdots}{c}\right)^2$ is actually finite, $\sim c^0$, rather than $\sim \frac{1}{c^2}$ as we naively assumed, because the singular roots of G_1 and G_2 are strongly correlated: the root of G_1 is $-\frac{1}{c} + O(1)$, while the two most singular roots of G_2 have exactly the same asymptotics: $-\frac{1}{c} + O(\sqrt{c})$: coefficients in front of $\frac{1}{c}$ are the same for all these three roots!

For $D(G_2)$ we have:

$$D(G_2) = d_2^2 r_{12} \sim (c^3)^{2 \cdot 6 - 2}\left[\left(\frac{\cdots}{c} - \frac{\cdots}{c}\right)\left(\frac{\cdots}{c} - \frac{\cdots}{\sqrt{c}}\right)^4\right.$$

$$\left. \times \left(\frac{\cdots}{\sqrt{c}} - \frac{\cdots}{\sqrt{c}}\right)\left(\frac{\cdots}{c} - \rho^{(2)}\right)^4\left(\frac{\cdots}{\sqrt{c}} - \rho^{(2)}\right)^4\left(\rho_1^{(2)} - \rho_2^{(2)}\right)\right]^2$$

$$\sim c^{30} \left(\frac{\cdots}{c} - \frac{\cdots}{c} \right)^2 \left(\frac{1}{c^{4+1/2+4+2}} \right)^2 (1 + O(c)) \sim c^9 \left(\frac{\cdots}{c} - \frac{\cdots}{c} \right)^2 (1 + O(c)).$$

We know already, that the remaining difference is not $\frac{1}{c}$, since both most singular roots of G_2 are $\sim -\frac{1}{c}$ with identical coefficients (-1). However, there is more: they are actually $-\frac{1}{c} \pm i\sqrt{c} + \ldots$, so that the difference is actually $\sim \sqrt{c}$, and $D(G_2) \sim c^{10}$. There is no similar cancellation between the two less singular roots, because they are actually $\sim \pm \frac{i}{\sqrt{c}}$ with *opposite* rather than equal coefficients. Thus, since we know asymptotics of $D(G_2)$ and r_{12}, $d_2^2 \sim c^5 (1 + O(c))$.

In this case the Mandelbrot set has no $\mathbf{Z_2}$ symmetry (like for the family $x^3 + c$), therefore we expect that all the cusps belong only to the zero-level components $\sigma_n^{(p)}$, i.e. $q(p) = 1$ for $p = 0$ and $q(p) = 0$ for $p > 0$ (like for the family $x^2 + c$). The sum rules (6.6) imply in this case:

$$\mathcal{N}_n^{(0)} + \sum_m \deg_c[r_{n,n/m}] = \deg_c[G_n(\varepsilon, c)],$$

$$\mathcal{N}_n^{(0)} = \deg_c[d_n],$$

$$\sum_{p \geq 1} \mathcal{N}_n^{(p)} = \sum_m \deg_c[r_{n,n/m}]$$

and the consistency condition

$$\deg_c[d_n] + \sum_m \deg_c[r_{n,n/m}] = \deg_c[G_n(\varepsilon, c)],$$

all obviously satisfied by the data in the table, provided the number of elementary domains is counted as the c-degree of $G_n(x, c)$ with small c-independent $x = \varepsilon$ (exactly at critical point $x = 0$ all $G_n(x, c)$ vanish, and at another critical point $x = -2/3c$ they have negative powers in c).

6.4.4 $f_\gamma = x^3 + \gamma x^2$

The family $cx^3 + x^2$ is diffeomorphic (i.e. equivalent) to $x^3 + \gamma x^2$ with $\gamma = 1/\sqrt{c}$, which is much simpler from the point of view of resultant analysis.

This time the map has two different critical points, $\{w_f\} = \{0, -2\gamma/3\}$. Therefore one could expect that the number of elementary domains is counted by a sum of two terms:

$$\# \text{ of elementary domains} \overset{?}{=} \deg_\gamma[G_n(0, \gamma)] + \deg_\gamma \left[G_n \left(-\frac{2\gamma}{3}, \gamma \right) \right].$$

However, $G_n(x = 0, \gamma)$ vanishes identically because of the high degeneracy of the map $f = x^3 + \gamma x^2$ at $x = 0$ and does not contribute. Also two out of three roots of $G_1\left(-\frac{2\gamma}{3}, \gamma\right)$ for $n = 1$ coincide (and equal zero), and label one and the same elementary domain. Thus, actually,

$$\text{\# of elementary domains} = \deg_\gamma\left[G_n\left(-\frac{2\gamma}{3}, \gamma\right)\right] - \delta_{n,1}.$$

$f = x^3 + \gamma x^2$					
n	1	2	3	4 ...	
# of orbits $= \deg_x[G_n]/n$	3	3	8	18	
# of el. domains= $\deg_\gamma[G_n\left(-\frac{2\gamma}{3},\gamma\right)] - \delta_{n,1}$	2	6	24	72	
d_n	$\gamma^2 + 4$	$i(4\gamma^4 - 13\gamma^2 + 32)$	*	*	
total # of cusps $= \deg_\gamma[d_n]$	2	4	20	62	
$r_{n,n/m}$ and total # of (n,n/m) touching points= $\deg_\gamma[r_{n,n/m}]$ for					
m=2	-	r_{12} $3\gamma^2 + 16$	-	r_{24} $20\gamma^6 + 47\gamma^4 - 220\gamma^2 + 1000$	
	-	2	-	6	
m=3	-	-	r_{13} $7\gamma^4 + 68\gamma^2 + 169$	-	
	-	-	4	-	
m=4	-	-	-	r_{14} $5\gamma^4 + 44\gamma^2 + 100$	
	-	-	-	4	
$\mathcal{N}_n^{(p)}$: # of el. domains $\sigma_n^{(p)} \subset S_n$ with q(p) cusps					
p=0, q=1	2	4 ?	20 ?	62 ?	
p=1, q=0	?	?	?	?	
p=2, q=0	?	?	?	?	
p=3, q=0	?	?	?	?	
...					

$$d_3 = 7168a^{20} + 42752a^{18} - 123072a^{16} + 233040a^{14}$$
$$+2585040a^{12} - 9493768a^{10} + 33146817a^8 - 107832976a^6$$
$$+315356704a^4 - 765891776a^2 + 1235663104$$

$$d_4 = 2319282339840\gamma^{62} + 99381248262144\gamma^{60} + 1782088231550976\gamma^{58}$$
$$+17132514753642496\gamma^{56} + 94391968968212480\gamma^{54} + 314124845396787200\gamma^{52}$$
$$+933600163325280256\gamma^{50} + 4359984836267474944\gamma^{48} + 16153160956010037248\gamma^{46}$$
$$+25910347327455363072\gamma^{44} + 19011408301729447936\gamma^{42} + 93048634733415038976\gamma^{40}$$
$$+165660797805589659648\gamma^{38} - 183114267457753161728\gamma^{36}$$
$$-295875264228762906624\gamma^{34} + 679482468842917743411 2\gamma^{32}$$
$$+338352329827825815024 64\gamma^{30} + 43548434837473605814592\gamma^{28}$$
$$+31982454515657753363720 0\gamma^{26} + 84722721943829250141604 8\gamma^{24}$$
$$-39135961449490819913564 0\gamma^{22} - 48445581970920710049171 19\gamma^{20}$$
$$-22393560229455792230880276\gamma^{18} + 44985877595412915098027 44\gamma^{16}$$
$$+318292811209086566680166 40\gamma^{14} + 571799816500523292950528000\gamma^{12}$$
$$-2515218215953699438592000 00\gamma^{10} + 1173757802691870851072000000 0\gamma^{8}$$
$$-1304466312653910835200000000 0\gamma^{6} + 8693002811987722240000000000 0\gamma^{4}$$
$$-3963167672086036480000000000 0\gamma^{2} + 180143985094819840000000000000$$

As already mentioned in the previous subsection 6.4.3, in this case the Mandelbrot set has no $\mathbf{Z_2}$ symmetry (like for the family $x^3 + c$), therefore we expect that all the cusps belong only to the zero-level components $\sigma_n^{(p)}$, i.e. $q(p) = 1$ for $p = 0$ and $q(p) = 0$ for $p > 0$ (like for the family $x^2 + c$). The sum rules (6.6) imply in this case:

$$\mathcal{N}_n^{(0)} + \sum_m \deg_\gamma[r_{n,n/m}] = \deg_\gamma\left[G_n\left(-\frac{2\gamma}{3},\gamma\right)\right] - \delta_{n,1},$$

$$\mathcal{N}_n^{(0)} = \deg_\gamma[d_n],$$

$$\sum_{p\geq 1}\mathcal{N}_n^{(p)} = \sum_m \deg_\gamma[r_{n,n/m}]$$

and the consistency condition

$$\deg_\gamma[d_n] + \sum_m \deg_\gamma[r_{n,n/m}] = \deg_\gamma\left[G_n\left(-\frac{2\gamma}{3},\gamma\right)\right] - \delta_{n,1},$$

all obviously satisfied by the data in the table.

Comparing the data in tables in this section and in the previous s.6.4.3 we see that the numbers of elementary domains differ by two – despite the two maps are equivalent (diffeomorphic) in the sense of Eq. (6.1). This is the phenomenon which we already studied at the level of quadratic maps (which are all diffeomorphic) in Sec. 6.3.5: different families of maps (even diffeomorphic) are different sections of the same entity – universal discriminant variety, but what is seen in the section depends on its (section's) particular shape. Relation between $cx^3 + x^2$ and $x^3 + \gamma x^2$ is such that $c = 1/\gamma^2$, so it is not a big surprise that in the γ plane we see each domain from the c plane twice.

It is instructive to see explicitly how it works, at least for the simplest case of $\sigma_1^{(0)}$. This (these) domain(s) is (are) the (parts of) stability domain S_1, defined by relations (4.19),

$$\begin{cases} cx^3 + x^2 - x = 0 \\ |3cx^2 + 2x| < 1 \end{cases}$$

for the family $cx^3 + x^2$ and

$$\begin{cases} x^3 + \gamma x^2 - x = 0 \\ |3x^2 + 2\gamma x| < 1 \end{cases}$$

for the family $x^3 + \gamma x^2$. The first of these systems is easily transformed to

$$\begin{cases} |3 - x| < 1, \text{ or } x = 3 - e^{i\phi} \\ c = \frac{1}{x^2} - \frac{1}{x} \end{cases}$$

while the second one – to

$$\begin{cases} |3 - \gamma x| < 1, \text{ or } x = \frac{3 - e^{i\phi}}{\gamma} \\ \gamma = \frac{1}{x} - x. \end{cases}$$

The critical point (single) of the first system, where $\partial c/\partial \phi = 0$, is at $x_{cr} = 2$, thus $c_{cr} = -\frac{1}{4}$ – the zero of discriminant $d_1(c) = 4c + 1$. The critical points of the second system, where $\partial \gamma/\partial \phi = 0$, are at $x_{cr} = \pm i$, thus $c_{cr} = -2 \mp i$ – the zeroes of discriminant $d_1(\gamma) = \gamma^2 + 4$. These points are the positions of the single cusp of a single elementary domain $\sigma_1^{(0)}$ in the first case and the two cusps of two elementary domains $\sigma_{1,\pm}^{(0)}$ (one cusp per domain) in the second case.

6.4.5   Map $f_{a;c} = ax^3 + (1-a)x^2 + c$

We now use another possibility, provided by the study of 2-parametric families: we look at the situation with broken $\mathbf{Z_d}$ symmetries and at the interpolation between $\mathbf{Z_d}$ and $\mathbf{Z_{d-1}}$. In the family $\{ax^3 + (1-a)x^2 + c\}$ the $\mathbf{Z_3}$ symmetry of Julia set and $\mathbf{Z_2}$ symmetry of Mandelbrot set occurs at $a = 1$, while at $a = 0$ it is reduced to $\mathbf{Z_2}$ for Julia sets and to nothing for Mandelbrot set. Accordingly, the number of cusps of elementary components of Mandelbrot set should decrease as a changes from 1 to 0. It is instructive to see, how continuous change of parameter a causes change of discrete characteristic, like the number of cusps (i.e. that of zeroes of irreducible discriminants $d_n(c)$). Part of the answer is given by Fig. 6.4: at $a = 1$ all the zero-level components $\sigma_n^{(0)}$ have the 2-cusp shape; when a is slightly smaller than 1, only remote components $\sigma_n^{(0)}$ change their shape from 2-cusp to 1-cusp type, while central domains stay in the 2-cusp shape; and the smaller a the closer comes the boundary between 2-cusp and 1-cusp shapes. Analytically, this means that positions of zeroes of $d_n(c)$ depend on a in a special way.

Transition reshuffling process ("perestroika") is a separate story, it is shown in Fig. 6.5 and it repeats itself with all the components on Mandelbrot set, though the moment of transition is different for different components, as clear from Fig. 6.4. Figure 6.5 demonstrates that what was a single central component M_1 with the multipliers-tree internal structure for the family $\{x^3 + c\}$ at $a = 1$, deforms and splits into many such components as a decreases from 1 to 0. Most of these detached components disappear at $c = \infty$ as $a \to 0$, but some remain, including the central domain M_1 and, say, the M_3-domain for the family $\{x^2 + c\}$. This picture clearly shows how the trail structure between M_1 and M_3 (and its continuation further, towards the end-point $c = -2$) is formed for $\{x^2 + c\}$ from a single domain M_1 for $\{x^3 + c\}$. Our analytic method allows to describe and analyze this process by tracking the motion of zeroes of resultants and discriminants with changing a. For our 2-parametric family[1]

$$d_1 = (1 + a)^2 + 2(a - 1)(a + 2)(2a + 1)c - 27a^2c^2,$$

$$r_{12} = a^5 \left[(a + 3)(3a + 1) - 2(a - 1)(a + 2)(2a + 1)c + 27a^2c^2 \right],$$

$$d_2 = a^{5/2}i \left[(4a^4 - 29a^3 + 82a^2 - 29a + 4) - 2a(a - 1)(a + 2)(2a + 1)c + 27a^3c^2 \right],$$

---

[1] Alternative technical approach (especially effective for a close to 1), can substitute our family by a diffeomorphic one, $\{x^3 + \gamma x^2 + \alpha\}$ with $\gamma = \frac{1-a}{\sqrt{a}}$, $\alpha = c\sqrt{a}$. For this family expressions for discriminants and resultants are somewhat simpler, for example

$$d_1 = -27\alpha^2 - (4\gamma^3 + 18\gamma)\alpha + (\gamma^2 + 4),$$

$$r_{12} = 27\alpha^2 + (4\gamma^3 + 18\gamma)\alpha + (3\gamma^2 + 16),$$

$$d_2 = i(27\alpha^2 + (4\gamma^3 + 18\gamma)\alpha + (4\gamma^4 - 13\gamma^2 + 32)),$$

$$r_{13} = 729\alpha^4 + (216\gamma^3 + 972\gamma)\alpha^3 + (16\gamma^6 + 144\gamma^4 + 351\gamma^2 + 27)\alpha^2$$
$$+ (4\gamma^5 + 22\gamma^3 18\gamma)\alpha + (7\gamma^4 + 68\gamma^2 + 169),$$

$$\cdots$$

Instead interpretation in the region of a near 0 is less transparent.

$$r_{13} = a^{22}(7 + 40a + 4c - 4a^5c + 48a^5c^2 + 15a^4c^2 + 216a^2c^3 + 16c^2$$
$$+ 40a^3 + 48ac^2 + 16a^6c^2 - 324a^4c^3 + 2ca - 216a^5c^3 + 75a^2 + 7a^4$$
$$+ 729a^4c^4 - 131a^3c^2 - 2a^4c - 8a^2c + 8a^3c + 15a^2c^2 + 324a^3c^3),$$

$$\cdots$$

$$w_{1,1} = \frac{27q^2 + 4p^3 + 24p^2 + 36p}{27} =$$
$$= \frac{4(2a^4 - 8a^3 + 3a^2 + 10a - 7) - 12a^2(a-1)^3c + 81a^4c^2}{81a^3}$$

$$\cdots$$

and positions of zeroes are shown in Fig. 6.6. As $a \to 0$, of two zeroes of $d_1(c)$ one tends to $\frac{1}{4} + O(a)$ (position of the cusp for the family $\{x^2 + c\}$), another grows as $-\frac{4}{27a^2} - \frac{2}{9a} - \frac{1}{36} + O(a)$; of two zeroes of $r_{12}(c)$ one tends to $-\frac{3}{4} + O(a)$ (position of the intersection $\sigma_1^{(0)} \cap \sigma_2^{(1)}$ for the family $\{x^2 + c\}$), another grows as $-\frac{4}{27a^2} - \frac{2}{9a} - \frac{1}{36} + 1 + O(a)$; both zeroes of $d_2(c)$ grow – as $-\frac{1}{a} + 2 + O(a)$ and $-\frac{4}{27a^2} + \frac{7}{9a} - \frac{16}{9} + O(a)$. On their way, the pairs of complex (at $a = 1$) roots of r_{12} and d_2 reach the real line Im $c = 0$ and merge on it when the higher discriminants

$$D(r_{12}|c) \sim (a-1)^2(a+2)^2(2a+1)^2 - 27a^2(a+3)(3a+1)$$
$$= 4a^6 + 12a^5 - 84a^4 - 296a^3 - 84a^2 + 12a + 4$$
$$= 4(a^2 - 5a + 1)(a^2 + 4a + 1)^2 = ((2a-5)^2 - 21)((a+2)^2 - 3)^2$$

and

$$D(d_2|c) \sim a^2(a-1)^2(a+2)^2(2a+1)^2$$
$$- 27a^3(4a^4 - 29a^3 + 82a^2 - 29a + 4) = 4a^2(a^2 - 8a + 1)^3$$

vanish. Actually, this happens at $a = \frac{5-\sqrt{21}}{2} \approx 0.208712152$ (i.e. $\gamma = \sqrt{3}$) and $a = 4 - \sqrt{15} \approx 0.127016654$ respectively. After that the roots diverge again and move separately along the real line towards their final values at $c = -\frac{3}{4}$ and $c = \infty$. We have similar behavior for zeroes of r_{13}, d_3 and of higher resultants. The "main" splitting, shown in the central picture (with $a = 1/7$) of Fig. 6.5, takes place at $a \approx 0.141$ – it corresponds to the splitting of "end-point" of the component M_1 of Mandelbrot set (which is a limiting point at $n \to \infty$ of the sequence of zeroes of resultants $r_{n,2n}$). The "final splitting", shown in the next picture (with $a = 1/8$) of Fig. 6.5, takes place at $a \approx 0.121$, which is the zero of the discriminant $D(w_{1,2}|c)$ – this is obvious from the fact that the "end-point" $c = -2$ of Mandelbrot set for $\{x^2 + c\}$ is a zero of $w_{1,2}(c)$, and two such end-points can be seen to touch at this cadre of Fig. 6.5. Of course, there is nothing special in $w_{1,2}$ – except for that 1 and 2 are small numbers and that $w_{1,2}(c)$ has real zeroes, – other stages of reshuffling (similar catastrophes, involving smaller components of Mandelbrot set) take place when other $D(w_{n,s}|c) = 0$.

From Fig. 6.6 it is clear, that for small a an exact – only mirror-reflected – copy of the Mandelbrot set for the family $\{x^2 + c\}$ (which is located near $c = 0$) is formed in the vicinity of $c = \infty$ (far to the left of Fig. 6.6). More evidence to this statement is provided by analysis of relation (4.19). For example, for stability domain S_1:

$$\begin{cases} |3ax^2 + 2(1-a)x| < 1, \\ c = x - (1-a)x^2 - ax^3. \end{cases}$$

At small a this domain decays into two identical parts:

$$\begin{cases} 2|x| < 1 + O(a), \\ c = x - x^2 + O(a), \end{cases}$$

and, for $x = -\frac{2(1-a)}{3a} - \tilde{x}$,

$$\begin{cases} \left| 2(1-a)\tilde{x}\left(1 + \frac{3a}{2(1-a)}\tilde{x}\right) \right| < 1, \\ c = x - (1-a)x^2 - ax^3 = \left(-\frac{4}{27a^2} - \frac{2}{9a} + \frac{2}{9}\right) - (\tilde{x} - \tilde{x}^2) + O(a). \end{cases}$$

In general, the map $x \to ax^3 + (1-a)x^2 + c$ is equivalent to $\tilde{x} \to a\tilde{x}^3 + (1-a)\tilde{x}^2 + \tilde{c}$ with

$$\tilde{c} = -c - \frac{4(1-a)^3}{27a^2} - \frac{2(1-a)}{3a} = \frac{2(a-1)(a+2)(2a+1)}{27a^2} - c.$$

This is the peculiar $\mathbf{Z_2}$ symmetry of the Mandelbrot set for our family.[2]

---

[2] To avoid possible confusion, note that $\mathbf{Z_2}$ symmetry, discussed in this paragraph, is not respected in Fig. 6.5, obtained with the help of *Fractal Explorer* program [10]. The program obviously misinterprets some properties of Mandelbrot sets in the case of non-trivial families, like $\{ax^3 + (1-a)x^2 + c\}$. Still it seems to adequately describe some qualitative features even for such families and we include Fig. 6.5 for illustrative purposes, despite it is not fully correct. Actually, the right parts of the pictures seem rather reliable, while the left parts should be obtained by reflection. Fully reliable is Fig. 6.6, but it is less detailed: it contains information only about several points of the Mandelbrot sets. Some of these points (located in the right parts of the pictures) are shown by arrows in both Figs. 6.5 and 6.6 to help comparing these pictures.

$f = ax^3 + (1-a)x^2 + c,$	small a [a near 1]				
n	1	2	3	4	...
# of orbits $= \deg_x[G_n]/n$	3	3	8	18	
# of el. domains $= 2\deg_c[G_n(0;c)]$	2 [1]	4 [2]	16 [8]	48 [24]	
total # of cusps $= \deg_c[d_n]$	2	2	12	40	
total # of (n,n/m) touching points $= \deg_c[r_{n,n/m}]$ for					
m=2	-	2	-	4	
m=3	-	-	4	-	
m=4	-	-	-	4	
$\mathcal{N}_n^{(p)}$: # of el. domains $\sigma_n^{(p)} \subset S_n$ with $q(p)$ cusps					
$p = 0,\ q = 1$ [q=2]	2 [1]	2 [0]	12 [4]	40 [16]	
$p = 1,\ q = 0$ [q=1]	0 [0]	2 [2]	4 [4]	? [4]	
$p = 2,\ q = 0$ [q=1]	0 [0]	0 [0]	0 [0]	? [4]	
$p = 3,\ q = 0$ [q=1]	0 [0]	0 [0]	0 [0]	0 [0]	
...					

Every map of our family has two critical points, $x = 0$ and $\tilde{x} = 0$, i.e.

$x = \frac{2(1-a)}{3a}$, which are both independent of c. Therefore the number of elementary domains, $\sum_{w_c} \deg_c[G_n(w_c; c)] = 2\deg_c[G_n(0; c)]$. In this table most data without square brackets corresponds to the case of small a, i.e. to the small deformation of the family $\{x^2 + c\}$. However, even a minor deformation causes immediate switch of degrees of $d_n(c)$ and $r_{n,k}(c)$ to their values, characteristic for the cubic map family $\{x^3 + c\}$. At the same time, the numbers $q(p)$ of cusps remain the same as they were for the quadratic map family $\{x^2 + c\}$. The table shows what this means for the numbers of elementary domains. Numbers in square brackets count elementary domains in the vicinity of the $\{x^3 + c\}$ family, when some of components got merged (see Fig. 6.5) and the numbers of cusps increased. Thus the table shows transition between cubic and quadratic families in terms of discrete numbers. Of course, at least one of the sum rules and consistency condition also change under the transition, so do the numbers $q(p)$:

$$\deg_c[d_n] = q(1)\deg_c[G_n(w_c, c)] + \mathcal{N}_n^{(0)},$$

and

$$\deg_c[d_n] + \sum_m \deg_c[r_{n,n/m}] = q(0)\deg_c[G_n(w_c, c)],$$

while two other relations remain the same for all values of a:

$$\mathcal{N}_n^{(0)} + \sum_m \deg_c[r_{n,n/m}] = \deg_c[G_n(w_c, c)],$$

$$\sum_{p \geq 1} \mathcal{N}_n^{(p)} = \sum_m \deg_c[r_{n,n/m}].$$

One more warning should be made: since transition occurs at different values of a for different components of Mandelbrot set (see Fig. 6.4), the meaning of "close" (when we mention a close to 1 or to 0) depends on considered component (actually, on the value of n in the table).

6.5 Quartic maps

6.5.1 *Map* $f_c = x^4 + c$

This section of Julia sheaf was used as an illustration in Figs. 3.1 and 3.2.
In this case

$$G_1 = x^4 - x + c,$$

$$G_2 = x^{12} + x^9 + 3cx^8 + x^6 + 2cx^5 + 3c^2x^4 + x^3 + cx^2 + c^2x + c^3 + 1,$$

$$G_{1,1} = G_2 - 1 = (x^4 + x + c)(x^8 + 2cx^4 + x^2 + c^2).$$

Factorization of the last polynomial is accidental and is lifted by infinitesimal variation of the family $x^4 + c$, e.g. provided by additional term px. The same happens with accidental factorizations in the following formulas. The triple critical point is $w = 0$.

$$w_1 = c^3,$$

$$w_{11} = c^6,$$

$$w_{12} = c^6(c^3 + 2)^3(c^6 + 2c^3 + 2)^3,$$

$$\cdots$$

$$
\begin{aligned}
d_3 = {}&(1099511627776c^{15} + 4367981740032c^{12} + 6678573481984c^9 \\
&+ 4811867160576c^6 + 1590250910976c^3 + 1312993546389) \\
&\cdot(18446744073709551616c^{24} + 14959156522273839 5136c^{21} \\
&+ 525955697232230481920c^{18} + 103853130552465948 6720c^{15} \\
&+ 1235398275557786386432c^{12} + 894978163534410547200c^9 \\
&+ 419572901219568058368c^6 + 181980149245232679936c^3 + 79779353642425058769)
\end{aligned}
$$

The sum rules (6.6) now imply:

$$\mathcal{N}_n^{(0)} + \sum_m \deg_c[r_{n,n/m}] = \deg_c[G_n(w_c, c)],$$

$$3\mathcal{N}_n^{(0)} + 2\sum_m \deg_c[r_{n,n/m}] = \deg_c[d_n],$$

$$\sum_{p \geq 1} \mathcal{N}_n^{(p)} = \sum_m \deg_c[r_{n,n/m}]$$

and the consistency condition

$$\deg_c[d_n] + \sum_m \deg_c[r_{n,n/m}] = 3\deg_c[G_n(w_c, c)],$$

all obviously satisfied by the data in the table.

$f = x^4 + c$					
n	1	2	3	4	$\cdots$
# of orbits $= \deg_x[G_n]/n$	4	6	20	60	
# of el. domains $= \deg_c[G_n(w_c,c)]$	1	3	15	60	
d_n	$-27 + 256c^3$	$65536c^6 + 152064c^3 + 91125$	*	?	
total # of cusps $= \deg_c[d_n]$	3	6	39	165	
$r_{n,n/m}$ and total # of $(n,n/m)$ touching points $= \deg_c[r_{n,n/m}]$ for m=2	-	r_{12} $256c^3 + 125$	-	r_{24} $\dfrac{16777216c^9 + 53411840c^6 + 59113216c^3 + 24137569}{9}$	
m=3	-	3	r_{13} $\dfrac{65536c^6 - 2304c^3 + 9261}{6}$	-	
m=4	- -	- -	- -	r_{14} $\dfrac{65536c^6 - 24064c^3 + 4913}{6}$	
$\cdots$	- -	- -	- -	- -	
$N_n^{(p)}$: # of el. domains $\sigma_n^{(p)} \subset S_n$ with $q(p)$ cusps p=0, q=3	1	0	9	45	
p=1, q=2	0	3	6	6	
p=2, q=2	0	0	0	9	
p=3, q=2	0	0	0	0	
$\cdots$					

6.6 Maps $f_{d;c}(x) = x^d + c$

Julia sets for these maps have symmetry $\mathbf{Z_d}$, because pre-image $f^{\circ(-1)}(x)$ of any point x is $\mathbf{Z_d}$-invariant set $\left\{ e^{\frac{2\pi i k}{d}} x^{(-1)}, \ k = 0, \ldots, d-1 \right\}$, and thus entire grand orbits, their limits and closures are $\mathbf{Z_d}$ invariant. Sections of Mandelbrot set, associated with these families, possess $\mathbf{Z_{d-1}}$ symmetry, because of invariance of the equations $G_n(x;c) = 0$ under the transformations $x \to e^{\frac{2\pi i}{d-1}} x, \ c \to e^{\frac{2\pi i}{d-1}} c$.

The critical point for all these maps is (multiple) $w_f = 0$. The values of polynomials G_n with $n > 1$ at this critical point, $G_n(0;c)$, depend on c^{d-1} only, therefore the systems of roots $G_n(0;c) = 0$ possess $\mathbf{Z_{d-1}}$ symmetry (and the root $c = 0$ of $G_1(0,c) = c$ is a stable point of this symmetry). This $\mathbf{Z_{d-1}}$ is the symmetry of entire Mandelbrot set.

$$G_1(0,c) = c,$$
$$G_2(0,c) = 1 + c^{d-1},$$
$$G_3(0,c) = 1 + c^{d-1}\left(1 + c^{d-1}\right)^d,$$
$$G_4(0,c) = 1 + c^{d-1}\frac{\left(1 + c^{d-1}\left(1 + c^{d-1}\right)^d\right)^d - 1}{1 + c^{d-1}}$$
$$= 1 + \sum_{s=1}^{d} \frac{d!}{s!(d-s)!} c^{(d-1)(s+1)}\left(1 + c^{d-1}\right)^{sd-1},$$
$$\ldots$$

Degrees of these polynomials satisfy recurrent relation

$$\deg_c[G_n(0,c)] = d^{n-1} - \sum_{\substack{k|n \\ k<n}}^{\tau(n)-1} \deg_c[G_k(0,c)],$$

which is similar to relation for powers of $G_n(x)$,

$$N_n(d) = d^n - \sum_{\substack{k|n \\ k<n}}^{\tau(n)-1} N_k(d).$$

Therefore $\deg_c[G_n(0,c)] = N_n(d)/d$.

Note that $F_n(f(0);c) = f(F_n(0;c)) - f(0)$, and if $F_n(0;c) = 0$, then $F_n(f(0);c) = 0$. For our maps $f(x) = x^d + c$, $f(0) = c$, and $F_n(0;c) = 0$ implies $F_n(c;c) = 0$, actually $F_n(c;c) = F_n^d(0;c)$.

The sum rules (6.6) imply:

$$\mathcal{N}_n^{(0)} + \sum_m \deg_c[r_{n,n/m}] = \deg_c[G_n(w_c,c)],$$

$$\deg_c[d_n] = (d-2)\deg_c[G_n(w_c,c)] + \mathcal{N}_n^{(0)},$$

$$\sum_{p\geq 1}\mathcal{N}_n^{(p)} = \sum_m \deg_c[r_{n,n/m}].$$

Consistency condition

$$\deg_c[d_n] + \sum_m \deg_c[r_{n,n/m}] = (d-1)\deg_c[G_n(w_c,c)].$$

All these relations are satisfied by the data in the table on p. 149.

$f = x^d + c$							
n	1	2	3	4	5	6	...
$N_n(d)$	d	$d(d-1)$	$d(d^2-1)$	$d^2(d^2-1)$	$d(d^4-1)$	$d(d^2-1)(d^3+d-1)$	
# of orbits $= \deg_x[G_n]/n = N_n(d)/n$	d	$\frac{d(d-1)}{2}$	$\frac{d(d^2-1)}{3}$	$\frac{d^2(d^2-1)}{4}$	$\frac{d(d^4-1)}{5}$	$\frac{d(d^2-1)(d^3+d-1)}{6}$	
# of el. domains $= \deg_c[G_n(w_c,c)] = N_n(d)/d$	1	$d-1$	d^2-1	$d(d^2-1)$	d^4-1	$(d^2-1)(d^3+d-1)$	
total # of cusps $= \deg_c[d_n]$	$d-1$	$(d-1)(d-2)$	$(d-1)(d^2-3)$	$(d^2-1)(d^2-d-1)$	$(d-1)(d^4-5)$	$(d-1)\cdot d^5 - 2d^2 - 3d + 2$	
total # of (n,n/m) touching points $= \deg_c[r_{n,n/m}]$ for							
m=2	-	$d-1$	-	$(d-1)^2$	-	$(d-1)^3 + 2(d-1)^2$ $= (d-1)^2(d+1)$	
m=3	-	-	$2(d-1)$	-	-	$2(a-1)^2$	
m=4	-	-	-	$2(d-1)$	-	-	
m=5	-	-	-	-	$4(d-1)$	-	
m=6	-	-	-	-	-	$2(d-1)$	
...							
$\mathcal{N}_n^{(p)}$: # of el. domains $\sigma_n^{(p)} \subset S_n$ with q(p) cusps							
p=0, q=d-1	1	0	$(d-1)^2$	$(d-1)^2(d+1)$	$(d-1)^2(d^2+2d+3)$	$(d-1)d(d^3+c^2-2)$	
p=1, q=d-2	0	$d-1$	$2(d-1)$	$2(d-1)$	$4(d-1)$	$2(d-1)+(d-1)^3$ $= (d-1)(d^2-2d+3)$	
p=2, q=d-2	0	0	0	$(d-1)^2$	0	$2(d-1)^2+2(d-1)^2$ $= 4(d-1)^2$	
p=3, q=d-2	0	0	0	0	0	0	
....							

Of the total of $(d-1)(d+1)(d^3+d-1)$ domains σ_6 there are:
- $(d-1)d(d^3+d^2-2)$ zero-level components $\sigma_6^{(0)}$ – the centers of new isolated domains $M_{6\alpha}$;
- $2(d-1)$ first-level components $\sigma_6^{(1)}$, attached to the single $\sigma_1^{(0)}$ at $2(d-1)$ zeroes of the resultant r_{16}, located on the boundary of $\sigma_1^{(0)}$, one between every of $d-1$ cusps and $d-1$ zeroes of r_{12} (where the $\sigma_2^{(1)}$ components are attached to $\sigma_1^{(0)}$);
- $(d-1)^3$ first-level components $\sigma_6^{(1)}$, with $d-1$ attached to each of the $(d-1)^2$ zero-level components $\sigma_3^{(0)}$ at $(d-1)^3$ of $(d-1)^2(d+1)$ zeroes of r_{36} in exactly the same way as $d-1$ components $\sigma_2^{(1)}$ are attached to the central $\sigma_1^{(0)}$ at zeroes of r_{12};
- $2(d-1)^2$ second-level components $\sigma_6^{(2)}$, with $2(d-1)$ attached to each of the $d-1$ first-level components $\sigma_2^{(1)}$ at zeroes of r_{26} in exactly the same way as $2(d-1)$ components $\sigma_3^{(1)}$ are attached to the central $\sigma_1^{(0)}$ at zeroes of r_{13};
- $2(d-1)^2$ second-level components $\sigma_6^{(2)}$, with $d-1$ attached to each of the $2(d-1)$ first-level components $\sigma_3^{(1)}$ at remaining $2(d-1)^2$ zeroes of r_{36} in exactly the same way as $2(d-1)$ components $\sigma_2^{(1)}$ are attached to the central $\sigma_1^{(0)}$ at zeroes of r_{12}.

For fragments of Julia sheaf for various families see Figs. 5.3 and 5.4 $(d=2)$, 6.8 $(d=3)$, 5.5 $(d=4)$, 6.9 $(d=5)$ and 6.10 $(d=8)$.

6.7   Generic maps of degree d, $f(x) = \sum_{i=0}^{d} \alpha_i x^i$

It is useful to introduce the following gradation:
$$x \to \lambda^{-1}x,$$
$$f(x) \to \lambda^{-1}f(x),$$
$$\alpha_i \to \lambda^{i-1}\alpha_i,$$
$$F_n(x) \to \lambda^{-1}F_n(x),$$
$$G_1(x) \to \lambda^{-1}G_1(x),$$
$$G_k(x) \to G_k(x) \quad \text{for } k > 1.$$

Then, since $G_k(x)$ is a polynomial of degree $N_k(d)$ in x, it is clear that
$$G_k(x) = \alpha_d^{\frac{N_k(d)-\delta_{k,1}}{d-1}} \prod_{l=1}^{N_k(d)} (x - \rho_l^{(k)})$$

where the roots $\rho_l \rightarrow \lambda^{-1}\rho_l$ are sophisticated algebraic functions of the coefficients α_i. Consequently, the resultant

$$R(G_n, G_k) = \alpha_d^{\frac{(N_n(d) - \delta_{n,1})N_k(d) + N_n(d)(N_k(d) - \delta_{k,1})}{d-1}} \prod_{l=1}^{N_n(d)} \prod_{l'=1}^{N_k(d)} (\rho_l^{(n)} - \rho_{l'}^{(k)}).$$

(6.12)

Assuming that $n > k$ we can simplify the expression for degree of α_d to $\frac{2N_n(d)N_k(d) - \delta_{k,1}N_n(d)}{d-1}$.

Example. As a simple application of these formulas, take $d = 2$. Then, as $a_2 \rightarrow 0$, all the roots grow as α_2^{-1}, $\rho_l^{(k)} \sim \alpha_2^{-1}$. The only exception is *one* of the two fixed points $\rho_+^{(1)}$ which remains finite, but its differences with other roots are still growing as α_2^{-1}. Then from (6.12) we conclude that in this case

$$R(G_n, G_k) \sim \alpha_2^{(2N_n(2)N_k(2) - \delta_{k,1}N_n(2))} \alpha_2^{-N_n(2)N_k(2)} = \alpha_2^{N_n(2)(N_k(2) - \delta_{k,1})}$$

(6.13)

We used this result in Sec. 6.3.3 above.

When $\alpha_d \rightarrow 0$ for $d > 2$, then $N_n(d-1)$ out of the $N_n(d)$ roots of $G_n(x)$ (points on the orbits of order n) remain finite, while the remaining $N_n(d) - N_n(d-1)$ grow as negative powers of α_d. However, different roots grow differently, moreover, some asymptotics can coincide and then differences of roots can grow slower than the roots themselves: all this makes analysis rather sophisticated (see subsection 6.4.3 for a simplest example).

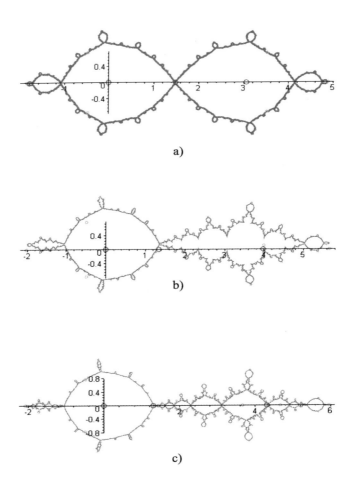

a)

b)

c)

Fig. 6.7   Coexisting stable orbits for the family $cx^3 + x^2$: a) $c = -0.22$, b) $c = -188$, c) $c = -0.178$. Julia sets are shown, and stable orbits lie inside $J(f)$. On the left-hand part of the Julia set there is a stable fixed point. The stable orbit inside the right-hand part is different: shown is the exchange of stability between *another* fixed point (stable in the case (a)) and an orbit of order two (stable in the case (c)). In (b) the transition point is shown, where the fixed point and the orbit of order two intersect and Julia set changes topology.

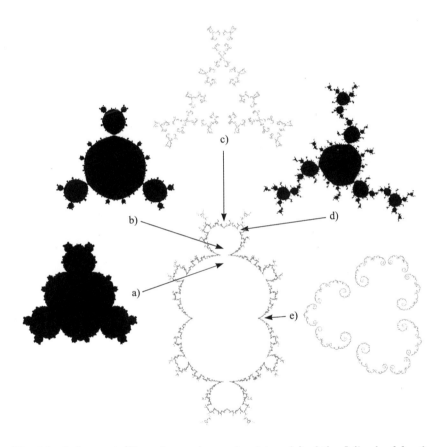

Fig. 6.8 A fragment (fibers shown at several points only) of the Julia sheaf for the family $x^3 + c$ – the analogue of Figs. 5.3, 5.4 for $x^2 + c$ and 5.5 for $x^4 + c$, a) $c = 0 + 0.7i$, b) $c = 0 + 0.9i$, c) $c = 0.2 + 1.08i$, d) $c = 1.14i$, e) $c = 0.45$.

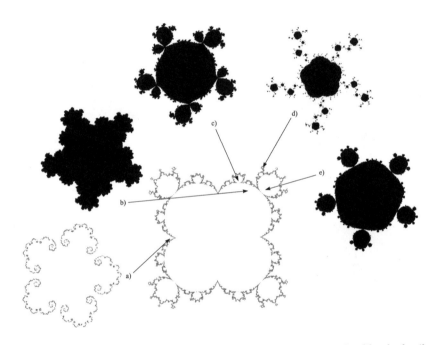

Fig. 6.9   A fragment (fibers shown at several points only) of the Julia sheaf for the family $x^5 + c$. Julia sets have symmetry $\mathbf{Z_5}$, Mandelbrot set has $\mathbf{Z_4}$ symmetry: a) $c = -0.6$, b) $c = 0.4 + 0.62i$, c) $c = 0.269 + 0.72i$, d) $c = 5827 + 0.8834i$, e) $c = 0.6 + 0.6i$.

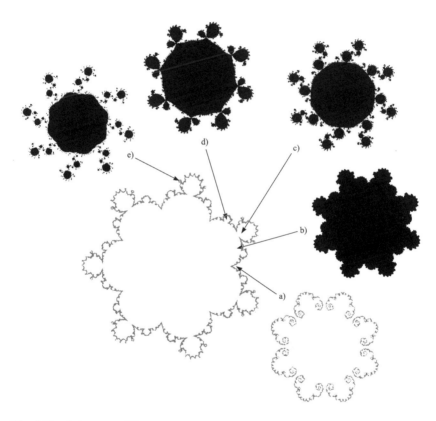

Fig. 6.10   A fragment (fibers shown at several points only) of the Julia sheaf for the family $x^8 + c$. Julia sets have symmetry $\mathbf{Z_8}$, Mandelbrot set has $\mathbf{Z_7}$ symmetry: a) $c = 0.7$, b) $c = 0.73 + 0.25i$, c) $c = 0.068 + 0.9775i$, d) $c = 0622 + 0.543i$, e) $c = 0.786 + 0.504i$.

Chapter 7

Conclusion

In this book we suggest to begin investigation of Julia and Mandelbrot sets as algebraic varieties in the phase space $\mathbf{X}$ and the moduli space of maps $\mathcal{M}$ respectively. We suggest to identify the pure topological and/or pure algebraic entity – the universal discriminant variety[1] $\mathcal{D}^* = \mathcal{R}^*$ of *real* codimension one in $\mathcal{M}$ – as the boundary ∂M of Universal Mandelbrot set M. The discriminant variety $\mathcal{R}^*$ naturally decomposes into strata of real codimension one,

$$\mathcal{R}^* = \bigcup_{k=1}^{\infty} \mathcal{R}_k^*$$

which, in turn, are made out of *complex*-codimension-one algebraic varieties $\mathcal{R}_{k,mk}^*$,

$$\mathcal{R}_k^* = \overline{\bigcup_{m=1}^{\infty} \mathcal{R}_{k,mk}^*}.$$

The intersection

$$\mathcal{R}_n^* \bigcap \mathcal{R}_k^* = \mathcal{R}_{n,k}^*$$

is non-empty only when k is divisor of n or vice versa. This internal structure of the universal discriminant variety is exhaustively represented by the *basic graph*: a directed graph, obtained by identification of all vertices of the multiples tree (see Sec. 2.3) with identical numbers at the vertices (this is nothing but the graph, discussed in Sec. 4.10 in relation to representation

---

[1]Since all resultants of iterated maps appear in decompositions of their discriminants, see Eq. (4.3), it is a matter of taste, which name, resultant or discriminant variety to use. However, to avoid confusion one should remember that elementary irreducible components are resultants $\mathcal{R}_{n,k}^*$, and irreducible discriminants are rare (they can be identified as regularized diagonal $\mathcal{R}_{nn}^*$.

theory of **Z**-action on periodic orbits). Irreducible resultants $\mathcal{R}^*_{k,mk}$ are associated with the links of this graph and $\mathcal{R}^*_k$'s – with its vertices. Each $\mathcal{R}^*_n$ separates the entire moduli space $\mathcal{M}$ into disjoint strata, with one or another set of periodic orbits of order n being stable within each stratum. An intuitive image can be provided by a chain of sausages: every sausage is stability domain for particular set of orbits of given order, every chain is bounded by particular $\mathcal{R}^*_k$, two chains, labeled by n and k touch along a spiralling line, represented by $\mathcal{R}^*_{n,k}$, see Fig. 7.1.

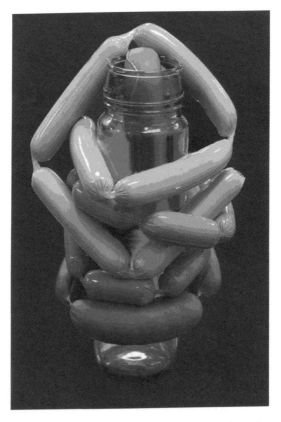

Fig. 7.1   A symbolic picture of the universal Mandelbrot set. A section can be obtained with the help of a knife.

Ordinary Mandelbrot sets, familiar from the literature, are obtained as sections of this universal discriminant variety by surfaces of low dimension, usually of complex dimension one. In such section a single stratum and a single irreducible resultant $\mathcal{R}^*_{n,mn}$ can be as a set of circles and points respectively (labeled by additional non-universal parameters α in Chap. 3). Irreducibility of resultants guarantee that all these points with given n and m belong to one and the same entity.

While such consideration seems to be potentially exhaustive for Mandelbrot sets (though their trail structure, shapes, fractal dimensions and Feigenbaum parameters still need to be better described in algebraic terms), it is not so clear about Julia sets: we did not identify them as (infinite) unions of well defined domains, and we did not establish full control over their geometry. Instead, see Sec. 5.5, we described Julia sets as strapped discs in $\mathbf{X}$, with identified points at the boundary. Identification of every group of points separates some sectors from the disc, and since at every bifurcation infinitely many groups of points are identified, the emerging structure of the Julia set looks fractal, see Fig. 7.2. Identified are points of various grand orbits, and orbits involved in this procedure depend on the place of the *map* in discriminant (Mandelbrot) variety: Julia sets form a kind of a sheaf over universal discriminant variety. We made just a few steps towards description of the structure of this sheaf, its monodromy and singularity properties. We came close to description of the underlying combinatorial structure, which is a sheaf of rooted trees (skeletons of Julia sets) over the basic graph (the skeleton of the discriminant/Mandelbrot variety).

Even more subtle is the issue of effective action and adequate description of cyclic and chaotic RG flows. Exact relation between τ-functions, phase transitions and Mandelbrot sets remains to be found. Generalizations to arbitrary fields $\mathbf{X}$ and possible relations to multi-dimensional flows are not explicitly worked out. Last, but not least, the obvious relations to **landscape theory**, which studies the distributions of algebro-geometric quantities on moduli spaces and their interplay with renormalization-group flows, are left beyond the scope of the present book. We are just at the beginning of an interesting story.

Fig. 7.2   Getting Julia set by strapping a ball. Actually there are double-infinity of strappings: going in the direction of particular phase transition we strap the points at the boundary, associated with different orbits; and even if a direction is fixed, infinitely many points, belonging to entire grand orbit, should be strapped.

Bibliography

[Arnold(1989)] For an introduction to the theory of dynamical systems see, for example:
  V.I.Arnold, *Mathematical Methods of Classical Mechanics*, (1989), Springer;
  H.Haken, *Synergetics, an Introduction* (1977) Springer, Berlin;
  R.L.Devaney, *An Introduction to Chaotic Dynamical Systems*, 2nd ed. (1989) Addison-Wesley Publ.;
  S.H.Strogatz, *Nonlinear Dynamics and Chaos*, (1994) Addison-Wesley Publ.

[1] see, for example, http://mathworld.wolfram.com/Characteristic.htm.

[2] P.F.Bedaque, H.-W.Hammer and U.van Kolck, *Phys.Rev.Lett.* **82** (1999) 463, nucl-th/9809025;
  D.Bernard and A.LeClair, *Phys.Lett.* **B512** (2001) 78, hep-th/0103096;
  S.D.Glazek and K.G.Wilson, *Phys.Rev.* **D47** (1993) 4657, hep-th/0203088;
  A.LeClair, J.M.Roman and G.Sierra, *Phys.Rev.* **B69** (2004) 20505, cond-mat/0211338; *Nucl.Phys.* **B675** (2003) 584-606, hep-th/0301042; **B700** (2004), 407-435, hep-th/0312141;
  E.Braaten and H.-W.Hammer, cond-mat/0303249.

[3] G.'t Hooft, *On Peculiarities and Pit Falls in Path Integrals*, hep-th/0208054.

[4] A.Morozov and A.Niemi, *Nucl.Phys.* **B666** (2003) 311-336, hep-th/0304178.

[5] M.Tierz, hep-th/0308121;
  E.Goldfain, *Chaos, Solitons and Fractals* (2004);
  S.Franco (MIT, LNS), Y.H.He, C.Herzog and J.Walcher, *Phys.Rev.* **D70** (2004) 046006, hep-th/0402120;
  J.I.Latorre, C.A. Lutken, E. Rico and G. Vidal, quant-ph/0404120;
  J.Gaite, *J.Phys.* **A37** (2004) 10409-10420, hep-th/0404212;
  T. Oliynyk, V. Suneeta and E. Woolgar, hep-th/0410001.

[6] For discussion of sophisticated branch/phase structure of effective actions see, for example,
  A.Alexandrov, A.Mironov and A.Morozov, *Int.J.Mod.Phys.* **A19** (2004) 4127-4165, hep-th/0310113; hep-th/0412099; hep-th/0412205.

[7] For basic ideas about the links between effective actions and integrability, between low-energy effective actions, generalized renormalization group and Whitham (quasiclassical) integrable dynamics see

A.Morozov, *Sov.Phys.Usp.* **35** (1992) 671-714 (*Usp.Fiz.Nauk* **162** 83-176), http://ellib.itep.ru/mathphys /people/morozov/92ufn-e1.ps & /92ufn-e2.ps; *ibidem* **37** (1994) 1 (**164** 3-62), hep-th/9303139; hep-th/9502091; A.Gorsky et al., *Nucl.Phys.* **B527** (1998) 690-716, hep-th/9802007.

[8] See, for example,
   J.Milnor, *Dynamics of one complex variable* (1991);
   S.Morosawa, Y.Nishimura, M.Taniguchi and T.Ueda, *Holomorphic dynamics* (2000) Camb.Univ.Press
   G.Shabat, *Lecture at Kiev School*, April-May 2002.

[9] For various descriptions of Julia and Mandelbrot sets see, for example,
   http://mathworld.wolfram.com/JuliaSet.html & /MandelbrotSet.html;
   R.Penrose, *The Emperor's New Mind* (1989) Oxford Univ.Press.

[10] A.Sirotinsky and O.Fedorenko, *Fractal Explorer*, http://www.eclectasy.com/Fractal-Explorer/index.html & http://fractals.da.ru.

[11] V.Dolotin, alg-geom/9511010.

[12] M.Feigenbaum, *J.Stat.Phys.* **19** (1978) 25, **21** (1979) 669;
   L.Landau and E.Lifshitz, *Hydrodynamics* (1986) Nauka, Moscow; sec.32.

[13] See, for example,
   S.Lang, *Algebra* (1965) Addison-Wesley Pub.Company;
   A.Kurosh, *Course of High Algebra* (1971) Moscow.
   For generalization to polylinear maps see [11].